DECARBONIZE THE WORLD

FRANK DALENE

DECARBONIZE THE WORLD

SOLVING THE CLIMATE CRISIS WHILE INCREASING PROFITS IN YOUR BUSINESS

Published by Advantage, Charleston, South Carolina.
Member of Advantage Media Group.

ADVANTAGE is a registered trademark, and the Advantage colophon is a trademark of Advantage Media Group, Inc.

Printed in the United States of America.

10 9 8 7 6 5 4 3 2 1

ISBN: 978-1-64225-274-3
LCCN: 2021916594

Cover design by David Taylor.
Layout design by Carly Blake.

I dedicate this book to our son, Kristofer Neal Dalene, who tragically died in an auto accident at twenty-one years of age. We will always miss you and remember you; you are my inspiration.

CONTENTS

Imagination Advances Knowledge

There are no walls in my mind. Don't put me in a box.

There are no fences in my thoughts. Don't inhibit me to roam.

There are no boundaries to my imagination. Don't end my dream.

There are no restrictions to my reasoning. Don't force me to conform.

Allow me to wander outside the box, to roam freely.

Allow my imagination to be endless, to reason unfettered thought.

No walls, no fences, no boundaries, no restrictions.

An open mind, unimpeded thought,

Infinite imagination and unrestricted reasoning

Will accomplish the unthinkable.

ACKNOWLEDGMENTS

Writing my first book appeared at first to be a daunting task. Without the following people it wouldn't have been possible.

First, I wish to thank my wonderful wife, Gwen, who for forty-three years has worked by my side in our businesses. It is because of her support and help at home and at work that I have been able to achieve everything in my life.

I wish to thank my brother and business partner, Roy, my better half, for his steadfast leadership both in our community and our businesses. Roy's wife, Lori, has also been a great help and support in our businesses and the extended family. We are very fortunate to have dedicated, hard-working employees and staff, allowing me to take the time to write this book.

There are many more people in my life I am thankful for who made this book possible; please accept my thanks. I wish to thank my entire team at Advantage who has done a terrific job helping me and guiding me through this process of writing a book. Without them, the book probably wouldn't have happened, certainly not at the level of quality they have helped achieve. I am very fortunate to have such a talented team. Thank you for helping me put this concept to paper.

I am truly blessed to have so many wonderful people in my life. Thank you. I wish to also thank everyone who picked up this book to read it. I'm grateful for your support by making these efforts worthwhile.

FOREWORD

Decarbonize the World cogently presents the fault lines and weakness of current policies and practices in mitigating and adapting to the increase in atmospheric carbon dioxide.

There is increasing awareness of the existential threat of climate change caused by anthropogenic greenhouse gasses, in particular from carbon dioxide. These are presented in scientific research measurements as well as supported by empirical evidence. The impacts on agriculture, transportation, water supply, homes, and, ominously, disasters and human security are increasingly reported by scientific research and the IPCC reports.

Recently, growing and expanding multidisciplinary and transdisciplinary studies on new and reemerging infectious diseases, increasing human health burdens, and mounting economic and financial costs link these with the rise in temperature and humidity.

The ecological, economic, financial, social, and political systems are interconnected and interacting. Changes caused in and by any component exacerbate changes in the other components and, importantly—on the whole—the human sphere.

The system and systemic megatransformations are unprecedented in scale and speed, and many are irreversible. There is not enough time for organisms, including human beings, to cope with and adapt to these changes.

The recent, record-breaking heat waves in Canada and the US western states and the droughts, forest fires, and floods are demanding response from the government and the private sector. Necessarily, there is growing recognition of the imperative need for people to reconsider their lifestyles and take commensurate responsibilities for personal stewardship of natural resources and protection of the environment.

The COVID-19 pandemic, continuing with its evolving variants, underscores the imperative for integrated and holistic policies, strategies, and practices. Critically, the pandemic also highlighted the need for cooperation and collaboration between and among institutions, sectors, and disciplines, as well as countries. International crisis requires international response.

With climate change, there are no vaccines or therapeutics that can cure and prevent the effects. Neither will face masks, washing hands, and social distancing be effective.

A CLEAN ENERGY REVOLUTION IS TAKING PLACE.

Globally, there are growing demands by the general public, in particular the youth, to move away from recycled rhetoric to actions and the necessity for bold, transformative, and innovative responses. Business as usual and more of the same approaches will neither be adequate nor work effectively.

A clean energy revolution is taking place.

At the April 2021 Climate Summit hosted by President Biden and attended by forty world leaders, the US pledged to cut carbon emissions by 50 to 52 percent below 2005 levels by the end of this decade, doubling the previous target. And at the June 2021 G7 Summit, a decarbonized future was endorsed. China, India, Indonesia, and the Republic of Korea are among an increasing number of countries that have also articulated plans for carbon-neutral energy.

In the US, President Biden is forcefully articulating a clean energy paradigm and underpinning this vision and leadership with the Build Back Better strategy that encompasses electric vehicles, carbon-neutral electricity, clean water, safe and good-paying jobs, and economic determinants in the comprehensive infrastructure plan for a fast transition toward a decarbonized future. An increasing number of countries are also initiating this paradigm change.

These are underscored by

- the US president's call for 100 percent clean electricity by 2035;
- the EU's objective to be climate neutral by 2050—an economy with net-zero greenhouse gas (GHG) emissions and aligning the EU's commitment to the Paris Agreement;
- Japan's Prime Minister Declaration that the country will strive for net-zero GHG emissions by 2050. The declaration has specific plans to fundamentally revise coal-fired power plants, promote R&D on second-generation solar photovoltaic technology and carbon recycling technologies, and establish a platform for national and subnational governments to discuss ways toward decarbonization; and
- China's 14th Five-Year Plan (2021–2025) for national economic and social development and the long-range objectives toward 2035 (14th FYP), which articulates the country's net-zero goals—the president's September 2020 address to the UN General Assembly stated that CO_2 emissions will peak before 2030 and carbon neutrality will be achieved before 2060.

Business, industry, and financial institutions are recognizing the unprecedented scale and speed of the mega changes that are driven and exacerbated by climate change and are taking major actions to decarbonize.

In December 2019, investors from around the world, representing some $37 trillion in assets, signed a letter calling on governments to step up their efforts against climate change.

US corporate leaders are responding to climate change with alacrity and have pledged to reduce greenhouse gas emissions. In April 2021, the We Mean Business Coalition, composed of 408 major businesses and investors, signed an open letter to President Biden stating their support for the administration's commitment to climate action and for setting a federal climate target to reduce emissions.

The motor vehicle industry will stop production of the internal combustion engine, after over a hundred years using this technology, an unprecedented technical revolution, with profound societal and product chain implications. General Motors announced it will make only electric vehicles by 2035, Ford will sell only electric vehicles in Europe by 2030, and 70 percent of Volkswagen's sales will be electric by 2030. Similarly, Jaguar plans to sell only electric cars from 2025 and Volvo from 2030. Japan is envisaged to continuing gasoline-electric hybrids, although Nissan was the first mass-market car manufacturer that globally launched the all-electric vehicle.

Financial lending is increasingly requiring an impact statement on climate for funding. In 2015, at the request of the G20 Finance Ministers and Central Bank Governors, the Financial Stability Board established the industry-led Task Force on Climate-Related Financial Disclosures.

An analysis of 215 of the world's largest corporations' disclosures by the CDP, an international not-for-profit charity headquartered in

London, found that these companies, unless they take proactive steps, could incur $1 trillion climate change–related costs. By the companies' own estimates, a majority of these financial risks could start to materialize in the next five years or so. However, many companies also see "moneymaking potential in climate change and estimated about $2.1 trillion of possible opportunities in a warming world, with the majority expected to materialize within the next five years."

The Geneva-based World Business Council for Sustainable Development July 2021 report "Net-Zero Buildings: Where Do We Stand?" analyzed six case studies and found:

- As much as 50 percent of whole life carbon emissions in a building come from embodied carbon (manufacturing of materials and the construction process), with most of it emitted immediately at the start of the life cycle.
- Typically as few as six materials account for 70 percent of the construction-related embodied carbon.

A mechanism that builds upon science and data, harnesses the positive attributes of market forces, is easy to understand and apply, addresses the integrated and holistic components of the product supply chain, and identifies the areas for increasing competitiveness and profitability of an enterprise will be a very important tool to address the existential crisis of climate change.

Maximizing the use of measurable and objective data and minimizing subjective information will encourage confidence in the mechanism and promote its greater use toward decarbonization. Such a mechanism is most timely.

ICEMAN is an excellent example of Praxis, turning vision, theories, concepts, and pioneering construction engineering experience in building energy-efficient homes into practice, and with the potential to identify and encourage innovations and inventions in all parts of the supply and product chain.

Nay Htun, PhD
Chemical Engineer
Former UN Assistant Secretary General, UNDP, UNEP
Program Director and Special Advisor, Business and Industry, UN Conference on Environment and Development (UNCED)
Adjunct Professor, Material Sciences and Chemical Engineering, Stony Brook University, State University of New York, New York
Fellow, Imperial College, London, United Kingdom

INTRODUCTION

In 2020, when much of the world went into lockdown due to the COVID-19 pandemic, it was devastating for just about every person on the planet. Economies buckled under the sudden shutdown of entire industries. Unemployment soared. We stopped traveling, stayed indoors. We stopped driving our cars, flying in airplanes, taking busses and trains. It was a year of fear and frustration and devastation.

But some saw a potential upside: with everyone staying home, maybe, just maybe, our greenhouse gas emissions would fall. Maybe, just maybe, we could slow the catastrophic momentum of the climate crisis.

Everyone thought that with the lockdowns across the globe, greenhouse gas emissions would fall dramatically in 2020. Sadly, this did not happen. According to the Research Division of the National Oceanic and Atmospheric Administration, carbon monoxide and methane emissions surged in 2020. In fact, carbon dioxide levels are higher now than they have been any time in the past 3.6 million years.[1]

1 NOAA Research, "Despite Pandemic Shutdowns, Carbon Dioxide and Methane Surged in 2020," April 7, 2021, https://research.noaa.gov/article/ArtMID/587/ArticleID/2742/Despite-pandemic-shutdowns-carbon-dioxide-and-methane-surged-in-2020.

This is the continuation of a trend that started in the Industrial Revolution and has increased exponentially ever since. There have been numerous international summits, protocols, and agreements that have attempted to establish mandates that will lower greenhouse gas emissions. But while emissions have fallen in some countries, including the United States and Europe, global emissions have continued to rise, thanks largely to surging emissions in China and India.

The most recent international agreement, the Paris Agreement, changed the way countries make commitments to combat climate change. Instead of putting forth a metric for nations to follow, as had been done in previous climate accords and agreements, the Paris Agreement asked each nation to set its own target.

WE ARE ALL PART OF THE SAME PLANET. WE ALL LIVE IN THE SAME CLOSED ENVIRONMENT. WE ALL HAVE THE SAME ATMOSPHERE. WE ALL BREATHE THE SAME AIR.

This may seem all well and good, but here's the problem: we are all part of the same planet. We all live in the same closed environment. We all have the same atmosphere. We all breathe the same air.

What happens in China impacts everyone in the world. What happens in the Amazon as they cut down the rain forest—which absorbs mass amounts of carbon dioxide—impacts the whole world.

We have to think globally. What happens there impacts what is happening here. And the good news is: what we do here can impact what is happening there.

We have to act locally. Change occurs at the grassroots level. When we take personal responsibility for our carbon emissions, it's infectious, and it can become viral in its own right. Consumers can change the world, influencing corporate decisions and behavior and thus making an impact on the global economic structure.

Right now, the international pressure, such as it is, is not incentivizing these countries to reduce their emissions expediently.

It's time to let the free market do the work.

Imagine a mechanism that harnesses the positive market forces of competitive advantage to reduce greenhouse gasses. Market forces can be devastatingly cruel and destructive. We saw that in the 2008 financial crisis. They are as powerful as forces of nature. They are unapologetic and undeniable. But what if we could harness them for good? What if we could use the power of market forces to reduce the global carbon footprint while simultaneously benefiting the economy?

Government leaders who create and enforce climate change mandates may not be able to implement policies strong enough to reach their desired goals. They may face fears from opposition leaders claiming their policies will cause economic harm or hardship to the poor and weak in society, which may lead to undesirable compromises.

Now, imagine a market-driven measurement system that is easy to comprehend and implement. That can potentially have far greater impact on reducing greenhouse gas emissions than any government-mandated policy could have. While a government may not easily or sufficiently be able to establish and enforce climate change mandates to reach their goals, a mechanism that harnesses positive market forces can be incredibly effective.

The market forces of competitive advantage already favor environmental responsibility. There is a green movement in this country, growing from the grassroots level. We have green homes being built, solar panels being installed, towns like my town of East Hampton setting goals to be 100 percent renewable.

It comes down to education and awareness. Once people become aware, they generally really want to do something.

Whether it's a cup of coffee, a pair of sneakers, or a car, consumers today want to know more and more about the products they're purchasing. Currently, they can find out where it was made, whether it's organic, and what it contains. But until now, there has been one thing they could not learn: the product's carbon footprint.

Now, imagine a mechanism that provides those consumers with an easy-to-understand measure of the greenhouse gasses emitted during the manufacturing process—the same way a box of Cheerios lists ingredients and nutrition facts. Imagine the impact the availability of this information could have on purchasing decisions.

Knowledge is futile if it is only present in the mind. It is pointless if it only exists in books. It is ineffective if it only resides in institutions. It is only when imagination finds an application for knowledge that it becomes useful.

Imagine being able to capture and apply knowledge about every product's carbon footprint in an attribute so simple, a consumer can understand it at a glance. Imagine the competitive advantage products with low carbon footprints would have, as consumers choose to purchase products that are better for the environment. Imagine how that will drive companies and manufacturers to lower their carbon footprints, and how that, in turn, would drive the creation of more renewable energy infrastructure, saving businesses money on their energy costs. Imagine how that would spread beyond the borders of this country, as manufacturers in other nations work to lower their carbon footprints in order to remain competitive.

Well, it's time to stop imagining. A mechanism is here that can accomplish all of this. I'd like to introduce you to ICEMAN.

ICEMAN—International Carbon Equivalent Mechanism Attributed to Neutrality—is an attribute based in science and mathematics by which consumers and buyers of products can evaluate the product's carbon footprint prior to purchase.

ICEMAN is an innovative mechanism that applies established sciences and protocols developed for the calculation of greenhouse gas emissions to calculate the sum of greenhouse gasses emitted throughout every part of the supply chain—for the product itself and for all the materials and components it contains. The calculation encompasses the taking of raw material from the ground, the entire manufacturing process, and the transportation of the product to market.

ICEMAN then mathematically translates this calculation into a percentage of carbon neutral, which can then be listed as a simple index number on a package. An index of fifty indicates that a product is 50 percent carbon neutral. An index of one hundred indicates that the product is 100 percent carbon neutral—that means its production had no negative impact on the environment.

ICEMAN is geared toward spreading knowledge at the individual level. Individual consumers everywhere in the world will be able to look at one simple number on the package of a product and know exactly what impact the production of that product has on the environment. No mystery, no subjective opinions on whether the product is "green" or not. Just simple, straightforward science- and math-based information.

By making a choice based on a favorable carbon-neutral attribute, consumers will be able to participate in a mechanism designed to reduce greenhouse gas emissions—participating more fully in the process of reducing global greenhouse gas emissions.

As green ways of thinking and purchasing continue to become more mainstream, the competitive advantage will go to those who are ahead of the curve in reducing their carbon footprint. This competitive advantage will lead to increased profits for those companies who take action to reduce their carbon footprint.

It's a supply and demand issue, not a policy issue. If consumers demand products with a lower carbon footprint, companies will work to meet that demand in order to stay competitive. While solutions like cap and trade are driven by policy, ICEMAN is not driven by policy, regulation, or mandates. It happens under its own steam, driven by the force of competitive advantage in the marketplace.

The fear is that any carbon tax like cap and trade or redistribution of wealth schemes like Carbon Fee and Dividend—the types of policies generally recommended by international consortiums like the Kyoto Protocol—will simply pass on the carbon costs to the consumer and negatively impact the economy, eventually causing inflation.

Adopting ICEMAN will have the opposite effect, increasing profit margins and reducing the costs of products at the same time as reducing carbon emissions. It's win-win. According to International Energy Agency data, the United States achieved the largest absolute reduction in carbon dioxide emissions of any country in 2019.[2] The US achieved this while continuing to build a robust economy. ICEMAN will continue to exponentially increase this trend of reducing greenhouse gas emissions while benefiting the economy.

Of course, none of this will matter if the carbon footprints of countries like China and India keep growing. But the good news continues: ICEMAN has the ability to impact the emissions created by the manufacturing of products globally.

The potential for ICEMAN to positively affect climate change is enormous and widespread. Not only will it motivate businesses to cut greenhouse gas emissions, but it will also have an effect at the national and global levels. States, regions, and even countries will be able to attract industries based on their low-carbon infrastructure. Develop-

---

2 International Energy Agency, "Global CO_2 Emissions in 2019—Analysis," February 11, 2020, https://www.iea.org/articles/global-co2-emissions-in-2019.

ing nations will have an incentive to decarbonize and build low-carbon-footprint infrastructures. Countries with high-carbon infrastructures will come under pressure to rebuild in order to prevent industries from moving to other lower-carbon locations. Those that regress or simply maintain the status quo may find themselves falling behind.

Climate change is one of the most—if not the most—pressing issues currently facing humanity. It is, without exaggeration, a matter of life and death; the future of the human race is at stake. ICEMAN is an actionable solution and model for reducing our carbon footprint, thereby preserving our planet and our ability to survive—all while benefiting businesses and the economy.

THE FUTURE OF THE HUMAN RACE IS AT STAKE.

If that sounds promising to you, let me share with you what exactly ICEMAN is, how it works, and why I believe it will be successful. The future is bright, and we can build it together.

CHAPTER 1

WHAT'S THE PROBLEM WITH THE WEATHER?

Here is the thing about climate change: it isn't something we need to believe in. It is objective. It is not a religion. It is a scientific fact based on information we have now. It simply is what it is.

The data is there. Our planet is changing in measurable ways. The amount of greenhouse gasses in our atmosphere is rising. The acidity levels in oceans and lakes around the world are rising. The temperature of the ocean is rising, and sea levels will therefore rise from thermal expansion alone, regardless of whether glaciers and ice caps are melting.

Wildlife is being decimated. We are seeing massive biodiversity degradation as entire species face endangerment and extinction. Lakes that once flourished with life have become barren wastelands—"dead lakes" that live up to their name. The oases that once dotted the planet's great deserts, providing sustenance to desert flora and fauna, have shrunk and vanished.

Annual CO_2 Emissions

Carbon dioxide (CO_2) emissions from the burning of fossil fuels for energy and cement production. Land use change is not included.

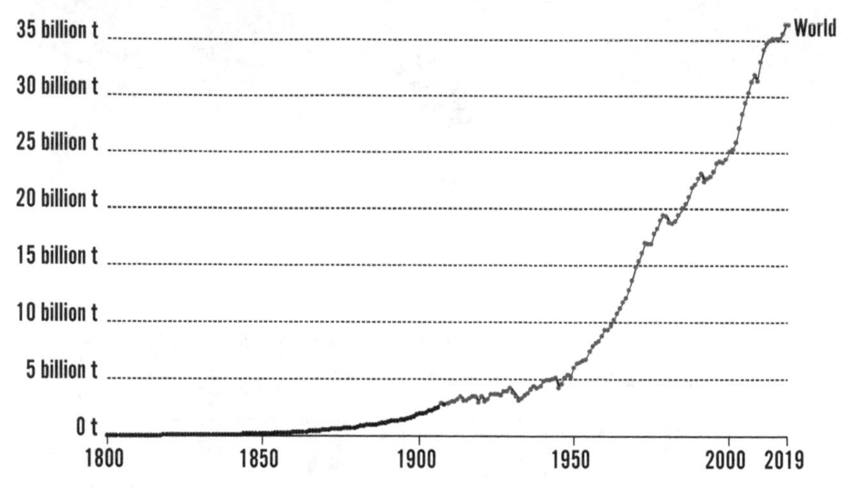

Note: CO_2 emissions are measured on a production basis, meaning they do not correct for emissions embedded in traded goods.

Source: Global Carbon Project; Carbon Dioxide Information Analysis Centre (CDIAC)

But it is not only wildlife under threat. The human species is in perilous danger as well. And not just some vague future danger. The crisis is already here. The weather is changing. Storms, hurricanes, tornadoes, and typhoons are all growing bigger, more powerful, and more frequent—as are floods, droughts, and wildfires. The destruction these natural disasters rain down upon the communities they hit is devastating, and the effects reverberate, as the ruination of farmland results in food shortages and famine.

All of these events and changes are measurable. There is no disputing the fact that they are, in fact, occurring. I always say that there's no such thing as a climate denier, because you cannot deny what is happening around us. There is hard evidence that climate change is occurring.

Shelf cloud leading impending thunderstorms, East Marion, New York.

Average Temperature Anomaly, Global

Global average land-sea temperature anomaly relative to the 1961-1990 average temperature.

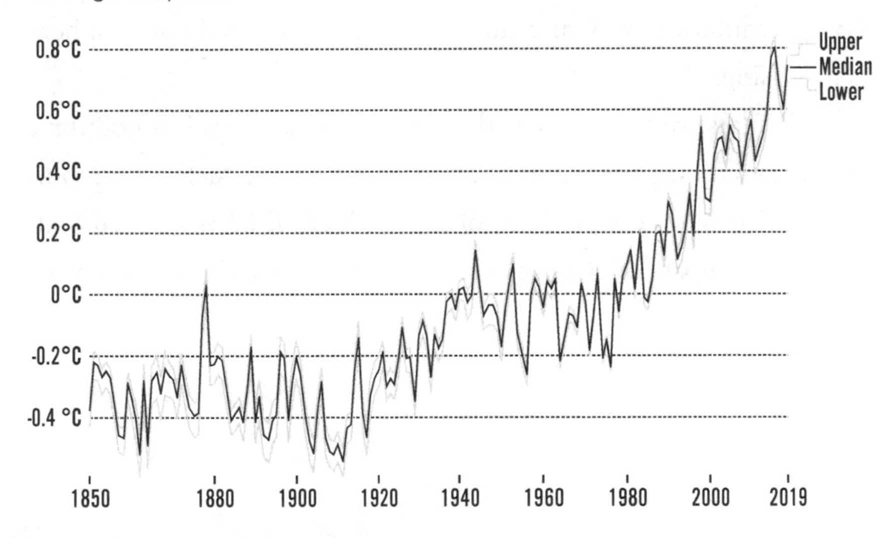

Note: The black line represents the median average temperature change, and grey lines represent the upper and lower 95% confidence intervals.

Source: Hadley Centre (HadCRUT4)

The true debate surrounding climate change is not about whether our environment is changing; rather, it is focused on the cause. Some argue that these changes are meteorological, that they simply represent the kind of long-term weather patterns our planet has always experienced. Once, we had an ice age; now, we are in an extended period of warming.

I am not disputing the existence of these long-term weather patterns. However, along with these patterns, we have measurable evidence that human activity has exponentially increased carbon emissions, releasing greenhouse gasses into our atmosphere. Since the Industrial Revolution, human beings have been bringing fossil fuels up from the ground and burning them for energy—in our cars, to create electricity, to heat and cool our water and air. When these fossil fuels are burned, they release pollutants into the air. Again, this has been observed and verified: the amount of CO2 and CO2 equivalents—the six types of greenhouse gasses that cause global warming (carbon dioxide, methane, nitrous oxide, hydro fluorocarbons, perfluorocarbons, and sulfur hexafluoride)—in the atmosphere is increasing.

The data we have now is that burning fossil fuels is a pollutant and that it is doing harm to the environment. We can all find common ground there. And we have a solution, which, if implemented, will reverse the impact of climate change while also benefiting both businesses and consumers.

Atmospheric CO_2 Concentration

Global average long-term atmospheric concentration of carbon dioxide (CO_2), measured in parts per million (ppm). Long-term trends in CO_2 concentrations can be measured at high-resolution using preserved air samples from ice cores.

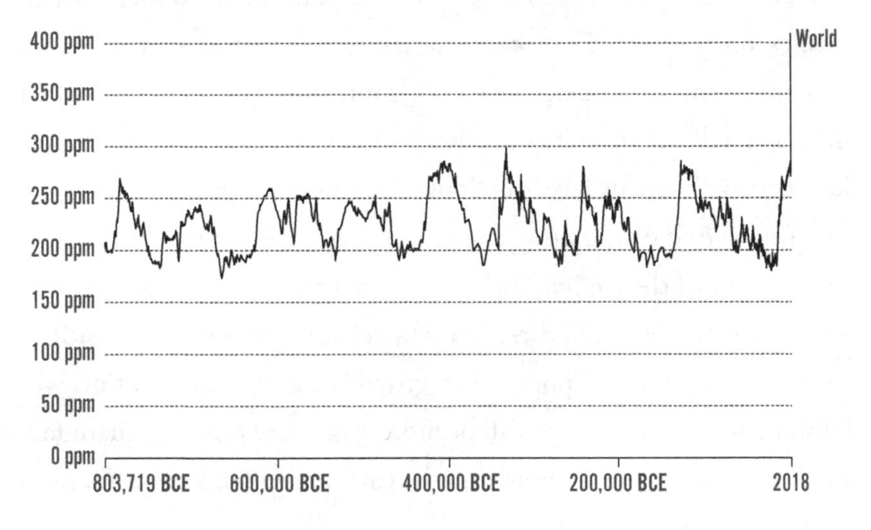

Source: EPICA Dome C CO_2 record (2015) & NOAA (2018)

WE HAVE TO ACT NOW

The truth is what we see on the surface is just the beginning of how bad climate change can get. The effects go much deeper, causing devastating damage and possibly catastrophic feedback loops. For example, as global temperatures increase, glaciers, polar ice caps, and the tundra's permafrost begin to melt. Hidden within the permafrost is methane. That methane has been trapped in the permafrost for hundreds of thousands, maybe millions of years. As the permafrost melts, the methane is released into the atmosphere.

Why is this such a big deal? Because methane is thirty-four times more powerful than CO_2 in creating climate change.[3] It is one of

3 Joshua Dean, "Methane, Climate Change, and Our Uncertain Future," EOS, May 11, 2018, https://eos.org/editors-vox/methane-climate-change-and-our-uncertain-future.

the greenhouse gasses that has the greatest impact. So, as methane is released from the permafrost, it contributes to rising temperatures. Increased temperatures cause more of the permafrost to melt, which releases more methane, which causes temperatures to rise more, so more permafrost melts, releasing more methane ... creating an ongoing feedback loop that accelerates the impact of climate change. Scientists believe this feedback loop will become irreversible.

This is one of the most dangerous feedback loops, but it is just one example of the feedback loops that are created by climate change. And we are perpetuating these loops by releasing more fossil fuels into the air. As the human population grows, human consumption will continue to increase. There will be more and more products manufactured, feeding more and more into the carbon feedback loops—unless we start to change things.

And we need to change things before it is too late—not for the planet, but for humankind. Throughout the history of the Earth, the planet has shown a remarkable ability to heal itself. My concern is not whether Planet Earth will survive; my concern is whether humans will survive this catastrophic effect.

We know both natural and manmade factors are at play. Our understanding of climate change and its causes will continue to develop as scientists continue to study the subject. These are the facts we have now. Something becomes "scientific fact" based on data and observations we know at a specific time. New data and discoveries may change the findings and determinations. Science evolves. That doesn't mean the science is wrong; it just means we have new data, and we adjust our findings and determinations. This has been the process since the beginning of scientific study. From Galileo to Newton to the leading scientists of today, science evolves and changes. The more we discover, the more we know. There's no ego involved; it is simply a matter of going on the data we have at the time.

Whatever the cause, the effect is the same: our environment has been severely impacted—and it's only getting worse. We can no longer sit idly by. Regardless of our individual position on causes, the time has come to step back, remove emotions, and take a logical view of what's happening. Logically, objectively, the moment of crisis is at hand. Something must be done, and the time to act is now.

SOMETHING MUST BE DONE, AND THE TIME TO ACT IS NOW.

WHY TOP-DOWN (ALONE) WON'T WORK

If the crisis is so dire, you may be thinking, Why hasn't the world been galvanized into action? Why hasn't anyone figured out a way to combat climate change that everyone can get behind?

Well, that's easier said than done.

Across the world—including in this country—there have been governmental efforts to reduce greenhouse gas emissions and reduce carbon footprints. The United Nations has negotiated and implemented several iterations of international environmental treaties under the United Nations Framework Convention on Climate Change, which was signed in 1992. First was the Kyoto Protocol in 1997, followed by the Paris Agreement in 2016.[4]

While many countries signed on to both treaties, the United States' adoption of the Kyoto Protocol was unanimously defeated in both the House and Senate. This unanimous decision was predicated

4 Lindsay Maizland, "Global Climate Agreements: Successes and Failures," Council on Foreign Relations, April 29, 2021, https://www.cfr.org/backgrounder/paris-global-climate-change-agreements.

in large part on a key provision of the Kyoto Protocol: the adoption of a cap-and-trade system.

Cap and trade is a system in which the government sets a cap on the amount of greenhouse gas emissions permitted across an entire industry. Then, companies within that industry trade in a market, buying and selling carbon offsets that permit them to emit a certain amount of greenhouse gasses. Because they need to pay for the carbon offsets, and because they can sell whatever offsets they don't use, companies have a strong incentive to reduce their emissions. Meanwhile, the government reduces the cap over time, encouraging companies to cut their emissions further. The government also sets the penalties for cap violations. It is a carrot-and-stick approach: the costs of purchasing carbon offsets and penalties for cap violations are passed onto the consumer, creating a carbon tax that increases the cost of products.

The Kyoto Agreement was also defeated because it included a poverty eradication element. The US was the wealthiest country with the highest per capita carbon footprint, so under the Kyoto Agreement it would have to pay fines, which would go to developing nations so they could start implementing renewable energy infrastructure. The UN and the countries that signed the agreement, most of which are social democracies, saw this as a reasonable arrangement. But in the US, it offended our capitalist, free market way of thinking and operating our economy.

Our solution has the opposite effect: it fits perfectly into our free market economy, incentivizing the use of renewable energy, creating a financial benefit for the company, and thus reducing the cost of products for consumers.

Cap and trade is not the only government-mandated system out there. Another proposed system for reducing carbon emissions that is currently working its way through Congress is the Carbon Fee and Dividend system. In this system, the government places a fee on fossil fuel emissions, which is steadily increased over time. This will increase the cost of products for companies, which will then increase the cost of those products for consumers. This is where dividends come in: the fees collected are placed in a trust fund, and the dividends for that trust fund are returned directly to consumer households every month. The argument goes that a majority of Americans would receive more in dividends than they would pay in higher prices.[5]

You'll notice that both of these systems start at the top—with the government. And this is precisely why they don't work in the United States, in a free market economy.

Many European countries, including my birth country of Norway, are social democracies, meaning they are governed from the top down. And for the most part, people in these countries trust their government. When I go to Norway and ask my family if they trust the government, they give me a strange look and say, "Of course, why wouldn't I?" In social democracies, people believe that the experts in the government, with their experience and knowledge, can determine what is best for the country and its citizens. Therefore, when the government passes a mandate like cap and trade, the people are more willing to accept that mandate.

The United States, from its birth, is structured in the exact opposite way. In the United States, power comes from the bottom up: from We, the People. It's how our constitution was written. From the Declaration of Independence onward, throughout American history, the power has always been with the people.

Now, I'm a realist. I know that there is centralized power in the

---

5 Citizens' Climate Lobby, "The Basics of Carbon Fee and Dividend," accessed February 3, 2021, https://citizensclimatelobby.org/basics-carbon-fee-dividend/.

government, and I know that the government doesn't always operate or act in exactly the way the people would like. But this country was founded on the concept that the people have the power. We elect our representative government. And our representatives are supposed to come back to the people and serve the people's will.

While the system doesn't always work perfectly, it remains true that in this country, many changes start at the grassroots level, with We, the People. When we have an agenda we want to push forward, we'll join a movement or an advocacy group. These movements and groups grow until there are enough people involved that our elected officials must listen to what we are requesting. We've seen movements gain momentum to the extent change becomes unstoppable. Consider the Women's Suffrage movement in the mid-nineteenth century and the Civil Rights movement led by Dr. Martin Luther King Jr.—movements that started with the people and changed how this country and our society operate.

Every change that happens in our country starts with social activists who lobby for a change they think is needed. When there is a big enough groundswell, the government begins to adopt that change—although usually not all of it and usually not all at once. Sometimes it happens overnight, but often it takes years and years.

Unfortunately, we don't have years and years. This crisis needs to be addressed immediately. That's why we need to harness the power that truly drives the American economy: market forces.

FROM THE BOTTOM UP: HARNESSING MARKET FORCES

A system like cap and trade can work, and work quickly, within a top-down social democracy. In a bottom-up free market economy—not so much. No mandatory system like cap and trade is going to effect real change in this country, and certainly not with the swiftness needed. In a bottom-up economy like the United States, a government cannot establish climate change mandates quickly and effectively enough to combat the exponential increase in greenhouse gas emissions. And they could never be enforced strongly enough; businesses would push back, and elected officials, worried about reelection, would refrain from pushing it harder. They would, as they are supposed to do, listen to the will of the people—but they can only do so much.

Government leaders who try to create and enforce climate change mandates are often faced by opponents who claim (sometimes erroneously, sometimes correctly) that their policies will cause economic harm, slow economic recovery, and cause hardship to the poor and underprivileged in our society. They argue that these are not compromises we can make.

But a climate change mechanism that can harness positive market forces can stimulate economic growth, corporate growth, and fiscal health—while also effectively combatting climate change—is a win-win situation for everyone, regardless of political belief or place in society. If we can harness the power of positive market forces, such as competitive advantage, we may actually be able to slow the impending—and indeed already occurring—climate crisis. In the United States, market forces may well be more powerful and effective in reducing greenhouse gasses than any government-mandated system.

Market forces are like forces of nature. They are incredibly powerful, and they can be absolutely brutal, cruel, and unforgiving. Market forces can destroy companies that are "too big to fail," like in the 2008 financial crisis. The largest banks, insurance companies, and auto manufacturers collapsed under negative market forces. The government had to step in and rescue these companies by infusing them with capital, like a rescue worker transfusing an accident victim with blood to save their life. That is how powerful these forces are. That is the level of destruction they can cause.

But market forces can also have an equally massive power for good. Like the forces of nature, market forces are what drive evolution and adaptation. What if we could harness the enormous power of market forces for a good cause? What if we could use the power of market forces to reduce greenhouse gasses? What if we could make reducing your carbon footprint a tangible competitive advantage in the marketplace?

The green movement has already gained momentum, thanks to growing consciousness of global warming, greenhouse gas emissions, and the consequences for our planet, our species, and our individual well-being. Consumers already prefer to purchase green products and services over those that are not, embracing products with green attributes. Corporations, businesses, manufacturers, and all producers of goods and services therefore already seek a competitive advantage by promoting the "green" attributes of their products or operations. The market forces are already at work. The fascinating realization is when these market forces are working, there is an added financial benefit that reduces costs, that goes down to the bottom line, increasing net profits or reducing the costs of products. In other words, they do the opposite of all currently known carbon-reducing systems.

I've seen this play out firsthand in the construction industry. In this industry, all sorts of energy-efficiency programs are being implemented and becoming part of the building code. While some are now required by the government, it all started with the US Green Building Council (USGBC)—another great illustration of a system growing up from the grassroots.

In the beginning, some of the founders of the USGBC stated that their goal was to create a system that could eventually become codified and used across the board. That hasn't happened yet, but the Energy Star Standard adopted by the US Environmental Protection Agency (EPA) and US Department of Energy (DOE) has become the adopted standard in our New York State energy building code. I assisted our New York State legislators to unanimously adopt enabling legislation for a property tax exclusion for USGBC LEED (Leadership in Energy and Environmental Design)–certified buildings in both the New York State Assembly and New York State Senate. And once again, I as a citizen brought this idea to our legislators. It wasn't the government that created this push for standardization; consumers creating the demand and businesses satisfying the demand are what did it, with no government interference at all. Quite the opposite, the government embraced and legislated what the people were already doing. It started as a voluntary program, and only when it achieved a certain level did it become law.

How did this come about? It wasn't just done out of care for the planet. It was also market-driven. Developers realized there was a competitive advantage to building USGBC-certified green buildings—both when it came to renting or selling apartments, condos, and offices and in reducing energy costs. And it has the health benefit of increasing indoor air quality. Studies proved that the indoor air quality of buildings that stuck to these environmental guidelines was markedly improved, and that, in turn, was conducive to more productivity at work and better concentration and less absenteeism in schools.

The data was there, and that made buildings that abided by these standards more desirable in the marketplace. Now, when businesses are renting an office, they would rather have a space with a LEED certification from the USGBC than an uncertified space. Big companies like PulteGroup, the third-largest home construction company in the country, are leading the way in adopting LEED certification. And LEED isn't the only certification offered by a nongovernmental organization; the National Association of Home Builders (NAHB) has its own certification as well, which is similar to but not exactly the same as USGBC's LEED certification.

All these certification systems have come up through the grassroots, and they're being adopted by major industry movers—not because of government mandates per se, but because there is demand in the marketplace, and the data proves it. A 2018 SmartMarket report on world green building trends found that green buildings have an asset value more than 10 percent higher than traditional, nongreen buildings.[6]

THE GREENWASHING CHALLENGE

The construction industry has reacted favorably to green certifications, demonstrating how market forces can support and move forward a green movement. And indeed, the green movement has gained a great deal of momentum in the market. Unfortunately, in its current form, it has also sowed seeds of doubt. Why? Because, so far, there have been no objective systems of measurement for certifying something as "green." The rating systems that exist currently have no scientific relationship to the greenness of the product. They are subjective, open to interpretation, which means certain green-rated products may contribute far less to protecting the environment than expected—or may not be green at all. This "greenwashing" has undermined the green movement.

6 Marisa Long, "Green Building Accelerates Around the World, Poised for Strong Growth by 2021," US Green Building Council, November 13, 2018, https://www.usgbc.org/articles/green-building-accelerates-around-world-poised-strong-growth-2021.

Even with industry organizations that have attempted to standardize green ratings through third-party certification programs like USGBC and NAHB, the problem persists. The points earned in both of those rating systems seem arbitrary. While the points may seem to align intuitively with green attributes, they have no direct scientific underpinning tying them to any actual measurements of a positive environmental impact.

Some green rating systems currently available have little objective relationship to how green a product actually is. In fact, they may even be entirely misleading—either unintentionally or purposefully. Manufacturers may be attaching green labels to products while their operations are spewing pollutants into the atmosphere—and we consumers would be none the wiser. A product may have a green label; it may even have some green attributes. But the operations of the manufacturer could actually cause more harm to the environment than any of the positive green claims.

Many so-called "green leaders" are no such thing. They tout themselves as environmentally friendly, plastering certifications and symbols all over their products and promotions. But in reality, these green accolades are self-assessed and self-awarded. They don't adhere to any actual objective standards. These companies claim to lead by example, but the example is empty of meaning.

The green label has been overused and abused. This hypocrisy erodes trust. People start to wonder: Is the whole green movement a sham? Does the green label actually have real meaning? Sadly, these practices have caused many people to become skeptical of anything "green," suspecting it of simply being greenwashing.

THE SOLUTION

So how do you regulate a grassroots movement that has grown from the bottom up? How can the "green" label be reclaimed, so that it carries actual objective meaning with scientific underpinnings?

We have an excellent example in another environmentally conscious movement: organic food. Organic food is a perfect illustration of the grassroots, ground-up process the green movement is currently experiencing. The organic food movement started at the grassroots level. In the 1950s and 1960s, organic gardening had some popularity, but it wasn't particularly mainstream. In the 1970s, in part due to the rise of the environmental movement, organic foods started to become more in demand. However, in order to determine whether your food was organic, you generally had to buy directly from growers. That was the only way to confirm that the food was not chemically treated or subjected to pesticides. It was up to the individual consumer to determine whether the food they were purchasing was organic.

In 1972, the International Federation of Organic Agricultural Movements was founded, with the intention of encouraging organic farming practices worldwide. Organic food continued to grow in popularity until 1990, and the organic food industry totaled around $1 billion in sales. This is when the US Congress passed the Organic Foods Production Act, which established uniform national standards for organic food. That Act authorized a new USDA National Organic Program (NOP) to certify that producers meet the standards set by the NOP. If producers meet the standards, they can label their products as "USDA Certified Organic."[7]

With this label, consumers could now see at a glance whether the

---

7 Statista, "Organic Food Sales in the U.S. 2019," May 2021, https://www.statista.com/statistics/196952/organic-food-sales-in-the-us-since-2000/.

product they were buying was organic, rather than having to determine that information for themselves. And they know they can trust that the product is, in fact, organic, even if they haven't connected directly with the producers of the food. Since then, organic food sales have continued to skyrocket. From 2005 to 2019, organic food sales in the United States grew from $13.26 billion to $50.07 billion.

Like the organic movement, the green movement started as a grassroots movement. Now, it has grown big enough that greenwashing has become a major issue, and some kind of objective, consistent standard needs to be applied.

Of course, organic food is just one industry. When we're talking about a system for certifying greenness based on the product's carbon footprint, we're talking about a system that can be applied to literally every industry and every product on the planet. How can we possibly create an objective standard that can be used across *every* industry, on *every* product? And how can that standard be simple enough that a consumer could actually understand it and take it into account in their purchasing decisions?

Enter ICEMAN.

Imagine an attribute, based in science and math, that could objectively evaluate a product or service's carbon footprint. Imagine that attribute to be easily communicable, so simple, that it can be put on a label. Imagine that a consumer can then look at that label and immediately understand that product or service's carbon-neutral status and can make a decision on what product or service to purchase factoring in that information.

This is exactly what the International Carbon Equivalent Mechanism Attributed to Neutrality (ICEMAN) offers. ICEMAN is an objective way to determine whether something is green. An ICEMAN label identifies, based on rigorous science and math, exactly how close a product or service comes to being carbon neutral. Why carbon neutral? This is the goal. When a product is carbon neutral, it means the manufacturing process and all materials or components no longer have a carbon footprint, meaning they no longer have an impact on climate change. The certified label would be able to identify not just the carbon footprint of the product itself but the complete carbon footprint of the manufacturer, including the overall embodied carbon footprint of the components and materials used in the product, all the way down the supply chain to when the raw material was taken out of the ground.

After the carbon footprint calculations are complete, a mathematical formula is applied to identify the percentage of a carbon-neutral status, known as the carbon index, the product achieved. The carbon index is a simple-to-understand 1–100 indexing system. Carbon neutral is the ultimate status every product should attain to eliminate carbon emissions in manufacturing, thus eliminating the climate change crisis caused by the carbon emissions of all manufactured products.

The ICEMAN index would restore integrity to the green movement by mathematically defining "green" based on measurements of greenhouse gas emissions, thereby objectively eliminating any kind of greenwashing or false claims of a carbon footprint being other than what it actually is.

This objective and quantifiable certification will be a powerful tool for companies to enhance their corporate image, promoting green attributes of their products or operations in a way their consumers can actually trust. The objective-, science-, and math-based ICEMAN certification will be impervious to greenwashing, thereby bolstering the positive image of the green movement.

Consumers have a growing awareness of the harmful effects of greenhouse gasses on our environment, and they tend to seek out products and companies that have made some kind of commitment to reducing their carbon footprint. Most businesses have realized by this point that there is a competitive advantage to marketing themselves as "green." That advantage will be much greater once manufacturers and businesses have the ability to demonstrate, scientifically and mathematically, the carbon neutrality of their product or service.

Knowledge is power. Giving consumers everywhere the ability to accurately and objectively evaluate products and services for their carbon footprints will have a potent impact on the market forces of competitive advantage. As consumers continue to embrace green products, a company that can deliver accurate, understandable, and reliable information will have an undeniable advantage.

When a consumer can quickly and clearly evaluate the carbon-neutral status of a product or service, they can easily consider that attribute alongside cost and quality when they make their purchasing decisions. This makes it easy to make an environmentally conscious choice, which is what most consumers want. A consumer will be more inclined to choose a product that is closer to carbon neutral. Therefore, manufacturers and businesses will have a strong incentive to create a product that has a higher carbon-neutral value, because that will give them a clear competitive advantage. As manufacturers and businesses create more carbon-neutral products, greenhouse gas emissions will be reduced faster and in greater amounts.

By virtue of making a choice based on favorable carbon-neutral standing, consumers will be able to participate in a mechanism designed to reduce the greenhouse gas emissions associated with manufacturing that product. And in the effort to reduce their carbon footprint, businesses and manufacturers will reduce their use of fossil fuels and start adopting renewable energy sources—which benefits all of humanity, and indeed the entire planet.

GOOD FOR BUSINESS, GOOD FOR THE WORLD

It doesn't matter whether you agree or disagree that the cause of climate change is manmade. We all agree the major source of carbon emissions is from fossil fuels: pollutants spewing particulates into the atmosphere causing major health problems, oil spills causing environmental disasters and many serious environmental consequences. The reduction of fossil fuels will cause vast environmental improvements around the world.

Our solution, if implemented, will benefit everybody. It won't hurt anyone. It won't tax anyone or raise the price of products.

Every carbon reduction solution presented so far includes some kind of tax, fee, or redistribution of wealth. Along with proposing mandates like cap and trade, the UN protocols also state that climate change solutions must include poverty eradication. Their plan to achieve this includes, in part, some redistribution of the United States' wealth to developing countries. This is another reason the US's adoption of the Kyoto Protocol was unanimously defeated in both the House and Senate.

ICEMAN, on the other hand, includes a financial benefit to the business, increasing their bottom line and saving them money, which can then lower the cost of the product to the consumer. It harnesses the power of knowledge and the power of the market forces of competitive advantage while diminishing carbon emissions.

So what exactly is ICEMAN, and how does it work? To explain, let me begin by sharing how ICEMAN came into being—which starts with my background in the construction industry.

CHAPTER 2

THE SEEDS OF ICEMAN

I was born in Norway. My family moved to the United States when I was eight months old, but I grew up in a Norwegian community, surrounded by Norwegian culture. In Norway, people spend a lot of time outside, hiking and walking and being out in nature. Because of this, Norwegians tend to have a strong consciousness of the environment. And so, in Norwegian culture, there is a huge emphasis on taking good care of our environment and being good stewards of our natural surroundings. This culture sparked a passion for the environment that has infused my whole career.

When I founded our building company, Telemark Inc., in 1978, I was already very aware of the early movement toward constructing energy-efficient houses. It wasn't called "green" back in those days; it was simply about energy efficiency. One of the first ads we ran highlighted our "energy-saving repairs." From the beginning, we adopted practices in our construction to make houses more energy efficient.

YUMPIN YIMMENY...Ven Vee Do A Yob Ve Do A Good Yob!

25 YEARS OF QUALITY CRAFTSMANSHIP WITH THAT NORWEGIAN TOUCH!

- Dormers
- Extensions
- Basements
- Kitchens
- Bathrooms
- Roofing/Siding
- Custom Building
- Any Alterations
- Energy Saving Repairs

TELEMARK CONSTRUCTION CO.
P.O. Box 186/220 Cambon Ave.
St. James, New York 11780

Licensed & Insured
Call for Realistic Estimates
516-472-4287
516-584-5151

Of course, those practices were a product of their time. For example, I worked with a concept called an "envelope house." An envelope house is designed with double walls, between which is a space. Within this space, air is circulated between the attic and the basement, running up and down the sides of the house. As the air in the southern walls is heated by the sun, it rises up one side of the house. As it moves through the attic, it cools down and drops down the other side of the house. This creates a natural convection of air around the entire building, which regulates the temperature in the house, saving energy on heating and cooling systems.

The concept never really caught on, because as we learned more about indoor air quality, we understood how much air exchange is needed to keep a healthy home. Having the same air constantly circulating isn't the healthiest system. Unclean air can cause not only allergies in those who are sensitive to allergens like dust and mold, but it can also carry bacteria that cause diseases like Legionnaires' disease or sick building syndrome. As an industry, we began to understand that we need air exchange to keep building environments healthy.

As a company, we were always on the leading edge of innovation, using the most cutting-edge heating and air-conditioning systems and even water source geothermal systems, which back in the early '90s were just coming into fruition. We were always early adopters of any technology that would save energy in our buildings. And in the Hamptons, where our business operated, people had the kind of budgets that allowed us to implement these innovations.

THE HGA'S AMBITIOUS GOALS

In 2008, I founded the Hamptons Green Alliance (HGA), an association of design professionals, building trades, manufacturers, suppliers, and related-service professionals organized to promote sustainable, resilient building and maintenance practices.

Around the same time, I received a newsletter from the Norwegian Embassy in Washington, DC, titled *News from Norway.* The cover story on one issue explained how Norway had committed to being carbon neutral by 2050.

"How can this be possible?" I asked myself. At the time, Norway was the third-largest oil exporter in the world. How could the third-largest oil exporter in the world go carbon neutral? My curiosity was piqued. I began studying the topic to learn how this was possible, which led me into a deep study of climate change science.

At an HGA meeting soon after, we were discussing what our first project as an association should be. We decided to send a letter to solicit the architectural community in search of an opportunity to build an ultra-green luxury home, but we wanted to set some specific goals for the project. Our first goal was to be USGBC LEED for

Homes Platinum—the highest level LEED certification. LEED is a widely used and internationally recognized green building certification system developed by the USGBC. The LEED for Homes rating system is based on points earned by meeting criteria in eight categories. There are four certification levels: Certified, Silver, Gold, and Platinum. Not only did we want to earn Platinum status, but we wanted to push the envelope and see just how many points we could get.

The second goal was to be net-zero energy. A net-zero energy building is a building that produces equal or more energy than it consumes, preferably through on-site, renewable sources. Fossil fuel–based energy production is minimized, generally using heat sources from solar thermal and geothermal energy production systems, along with renewable photovoltaic systems, such as solar roof panels, and wind generators. The building usually remains attached to the electric grid with a net-meter installed to put extra electricity produced by the solar panels on the grid during the day when the sun is bright and the occupants are at work and to take electricity from the grid at night when the sun isn't shining and the solar panels aren't producing electricity.

I FELT LIKE THERE WAS SOMETHING ELSE WE COULD BE DOING.

The Association agreed that a LEED Platinum certification and net-zero energy were excellent goals. But I felt like there was something else we could be doing.

"It should be carbon neutral!" I blurted out. I hadn't been able to stop thinking about going carbon neutral since reading that newsletter, and I knew it should be part of our project. I wanted to see if we could build a carbon-neutral house. So carbon neutral became our third goal.

WHAT IS CARBON NEUTRAL?

What does it mean to be carbon neutral? In the simplest terms, it means you are not leaving a carbon footprint. A carbon footprint is the total amount of greenhouse gas emissions caused by any given person, product, process, or activity. You can even find out your carbon footprint as an individual. There are a number of organizations with tools online, where you can put in your energy usage, what kind of car you have, how much gas you use, how much you travel, etc., as well as all your offsets, and can certify yourself as carbon neutral.

A carbon footprint begins as soon as a natural resource is pulled from the earth, and as it continues on its path to production—as each resource is joined with other resources to make a product—the carbon footprint expands.

The majority of greenhouse gas emissions—about 75 percent—comes from the burning of fossil fuels.[8] Wherever energy involving fossil fuels is used, it is a large part of what is measured when measuring carbon footprints. Any energy used that comes from renewable sources—solar, wind, hydro, biogas—is not part of a carbon footprint.

Greenhouse gasses are also given off by the manufacturing of certain materials (i.e., in the production of that material itself, in addition to the greenhouse gasses given off by energy used to create it). For instance, a lot of energy is used in creating concrete, because the limestone has to be heated up to a very high temperature. The process of producing clinkers from limestone emits greenhouse gasses in addition to the energy used. This information is tracked and catalogued in databases kept by government organizations, including the EPA and the DOE, as well as organizations in other countries. You can

8 US Energy Information Administration, "Where Greenhouse Gases Come From," August 11, 2020, https://www.eia.gov/energyexplained/energy-and-the-environment/where-greenhouse-gases-come-from.php.

get lists on these organizations' websites of the carbon footprints of different materials, and even an order of magnitude of which materials have the biggest carbon footprints. Concrete has one of the highest carbon footprints when it comes to construction material; aluminum is the highest.

When evaluating a business's carbon footprint, the past year of activity is taken into account. So when you are certified carbon neutral, it means whatever activities or projects you are working on that year do not leave a carbon footprint.

In order to be carbon neutral, you mitigate your carbon emissions as much as you can through energy efficiency measures and other environmentally conscious choices. Of course, it is generally not possible to completely eliminate all greenhouse gas emissions. So whatever you can't mitigate, you offset.

A carbon offset is a reduction in emissions made to compensate for carbon emissions produced somewhere else. This could be within the same operation—for example, producing renewable energy in one part of your operation in excess of what energy you consume in order to make up for carbon emissions related to another part of your operation.

Or you can purchase a carbon offset credit. There are a number of certification programs that certify projects that take greenhouse gasses out of the atmosphere, generating carbon offset credits based on how much greenhouse gas the projects remove. A buyer can then purchase carbon offset credits equal to the amount of greenhouse gas emissions their operation produces. The purchasing of these credits funds the projects themselves.

Trading offsets like this is part of how cap and trade operates. Companies that produce greenhouse gasses beyond the allowed cap for their industry must purchase offsets. That then funds projects that remove greenhouse gasses from the atmosphere. Carbon offsets are traded on an exchange or can be purchased in the private market.

When the total GHG emissions are offset 100 percent by carbon mitigation offsets and purchased carbon offsets, carbon-neutral status is achieved.

A NEW FRONTIER

After the HGA announced our goals for the house, I set out in search of a methodology or organization that would help me measure the carbon footprint of our construction—to help us achieve our carbon-neutral goal—but found none.

Measuring the carbon footprint of a building is broken down into three stages: construction, operations, and demolition. In the life cycle of a building, the operations stage covers the life span of the building, the length of time it is in use. It could be thirty years, fifty years, or longer. The carbon footprint of a building's operation stage means the total energy the building uses throughout its entire life span.

This accounts for 80 percent of the total carbon footprint of the building—and that's why I couldn't find anything about construction. I looked at the World Resources Institute (WRI), the United Nations Environmental Program (UNEP), and the Sustainable Buildings and Climate Initiative (SBCI), formerly the Sustainable Buildings and Construction Initiative, which is a partnership between the private sector, government, nongovernment, and research organizations formed to promote sustainable building and construction globally. All these organizations were working on measuring the carbon footprint of the operations stage through a methodology known as the Common Carbon Metric. No one was doing work on the construction stage—on all the energy used in making the materials, transporting them, construction, plus the carbon footprint of every subcontractor's business.

In other words, no one was working on the analysis of the entire life cycle of buildings. Life cycle analysis (LCA) is the accounting of greenhouse gas emissions during a product's entire life cycle, from the gathering of raw materials to disposal at the end of the product's life. Cradle-to-gate LCA is the accounting for greenhouse gas emissions from the beginning of a product's life until it reaches the consumer—in terms of buildings, that would be the construction stage. Determining the LCA data on building materials, products, and facilities is still in its infancy, with only a few private companies that voluntarily perform LCA on their products. LCA standards created by the WRI are still in draft form, and little information was currently available.

So, I had to create my own methodology.

Thankfully, this did not mean starting completely from scratch. It simply meant applying existing protocols based on accepted science to the construction phase. To develop a methodology for measuring the carbon footprint of the construction phase, I applied the WRI Greenhouse Gas Protocol.

THE WRI GREENHOUSE GAS PROTOCOL

The WRI Greenhouse Gas Protocol breaks emissions down into two categories: direct and indirect. Direct emissions are emissions from sources that are owned or controlled by the company. Indirect emissions are emissions from sources not controlled or owned by the company but are produced as a consequence of the company's activities—for example, the production of electricity.

There are four main sources of direct and indirect emissions:

stationary combustion, which is the burning of fuels in stationary equipment—boilers, furnaces, engines, etc.; mobile combustion, which is the emissions from transportation devices including cars, trucks, trains, ships, airplanes, buses, etc.; process emissions, which are emissions from processes such as the manufacturing of cement or aluminum smelting; and fugitive emissions, which are intentional or unintentional releases from equipment leaks, cooling towers, etc. The WRI Greenhouse Gas Protocol website offers calculation tools for all these sources.

The protocol then breaks these categories down into three scopes.

Scope 1 is all direct greenhouse gas emissions. This includes any emissions from vehicles, furnaces, equipment, machinery, etc. owned and/or controlled by the company. Any emissions created by the generation of electricity, heat, or steam through the burning of fossil fuels in boilers, furnaces, etc. owned by the company are included in Scope 1. So, for example, the greenhouse gas emissions from the oil that is being burned in your furnace to create heat would be considered Scope 1.

Also included in Scope 1 are emissions created by the manufacture or processing of materials or chemicals, such as cement or aluminum, by the company. Scope 1 also includes the emissions from all transportation of the following: materials, products, waste, and employees by vehicles owned by the company. Any emissions from a vehicle owned or controlled by the company for transportation are part of Scope 1. This includes any cars or trucks used by the company, as well as any other transportation services owned or controlled by the company—trains, ships, airplanes, buses, etc.

Scope 2 is indirect greenhouse gas emissions from the purchased electricity consumed by the company (i.e., the electricity the company has paid for, not any electricity produced by the company itself). This electricity is indirect, because it is produced offsite.

This includes all the electricity used by all parts of the company—equipment, buildings, etc. This accounts for the majority of indirect emissions for most companies. There are many energy-efficient technologies out there that can help a company reduce their electricity usage, and that is an important way companies can reduce their carbon footprint.

Scope 2 greenhouse gas emissions are dependent on what grid you are drawing your energy from. Different electrical grids in different locations have different carbon footprints, depending on the mixture of fuels used to produce electricity on that grid. The EPA calculates the energy that is used in producing electricity in that area and lists that information on their website. I can look up the Long Island Electric Grid on the EPA website, and it will give me the carbon emissions per kilowatt hour. On the Long Island grid, we have fuel oil and natural gas, as well as renewable hydroelectric energy imported from Niagara. In Tennessee or Kentucky, you might have more coal plants. Coal is dirtier than oil or natural gas, so producing electricity would have a higher carbon footprint in those areas than in areas that use oil or natural gas or renewable energy.

Scope 3 is all other indirect greenhouse gas emissions. Scope 3 is broken down into upstream and downstream emissions.

Upstream emissions are the Scope 3 emissions that occur "cradle-to-gate"—from inception until the product or service leaves the company to go into the hands of consumers. Upstream emissions include all emissions from transportation not completed by company-owned vehicles. That includes any transportation of materials, products, and fuel purchased by the company, as well as all waste generated by the company, that are not transported by company-owned vehicles. It also includes all transportation of employees: employees commuting to and from work, whether in their own cars or on public transportation, any business travel undertaken by employees, etc. Upstream emissions also include any emissions from the disposal and treatment of waste generated by the company's operations.

Upstream emissions also include all leased assets, franchises, and outsourced activities—so it includes all the emissions from any activities by any subcontractors done for the company, as well as any emissions produced by any assets leased by the company.

This also applies for all products and services purchased or acquired by the company. Upstream emissions include all upstream ("cradle-to-gate") emissions from the production of any product or service purchased or acquired by the company. This also includes the emissions created by the production of purchased materials. For example, if the company purchases concrete, the emissions from the production of that concrete is included in Scope 3. If the company purchases appliances to put in a house they are constructing, the upstream emissions from the creation of those appliances are included.

Downstream emissions are the emissions that occur once the product or service leaves the company. This includes all the transportation and distribution of products (that are not completed using company-owned vehicles or facilities), as well as the processing, use, and end-of-life treatment of sold products. Once the product or service is in the hands of the consumer, the carbon emissions are considered "downstream."

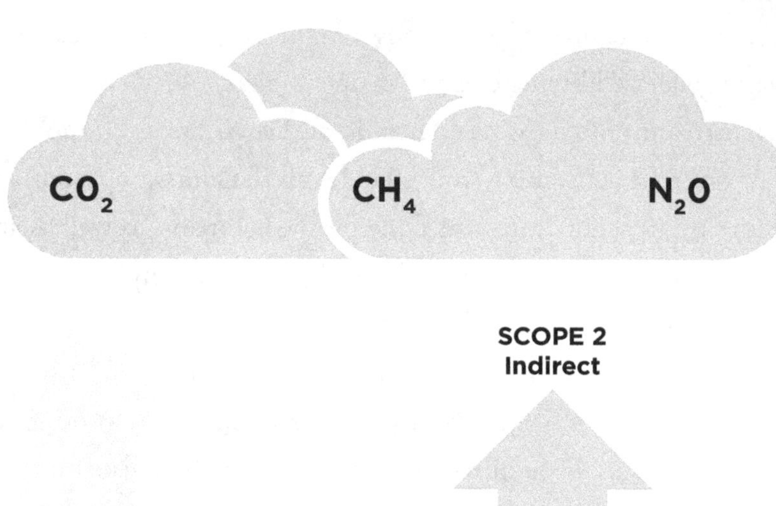
CO_2
CH_4
N_2O

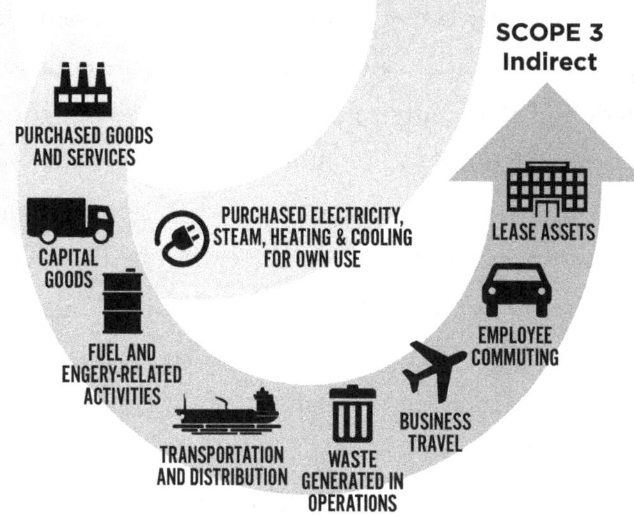
SCOPE 2
Indirect
SCOPE 3
Indirect
PURCHASED GOODS AND SERVICES
CAPITAL GOODS
FUEL AND ENGERY-RELATED ACTIVITIES
TRANSPORTATION AND DISTRIBUTION
WASTE GENERATED IN OPERATIONS
BUSINESS TRAVEL
EMPLOYEE COMMUTING
LEASE ASSETS
PURCHASED ELECTRICITY, STEAM, HEATING & COOLING FOR OWN USE

UPSTREAM
ACTIVITIES

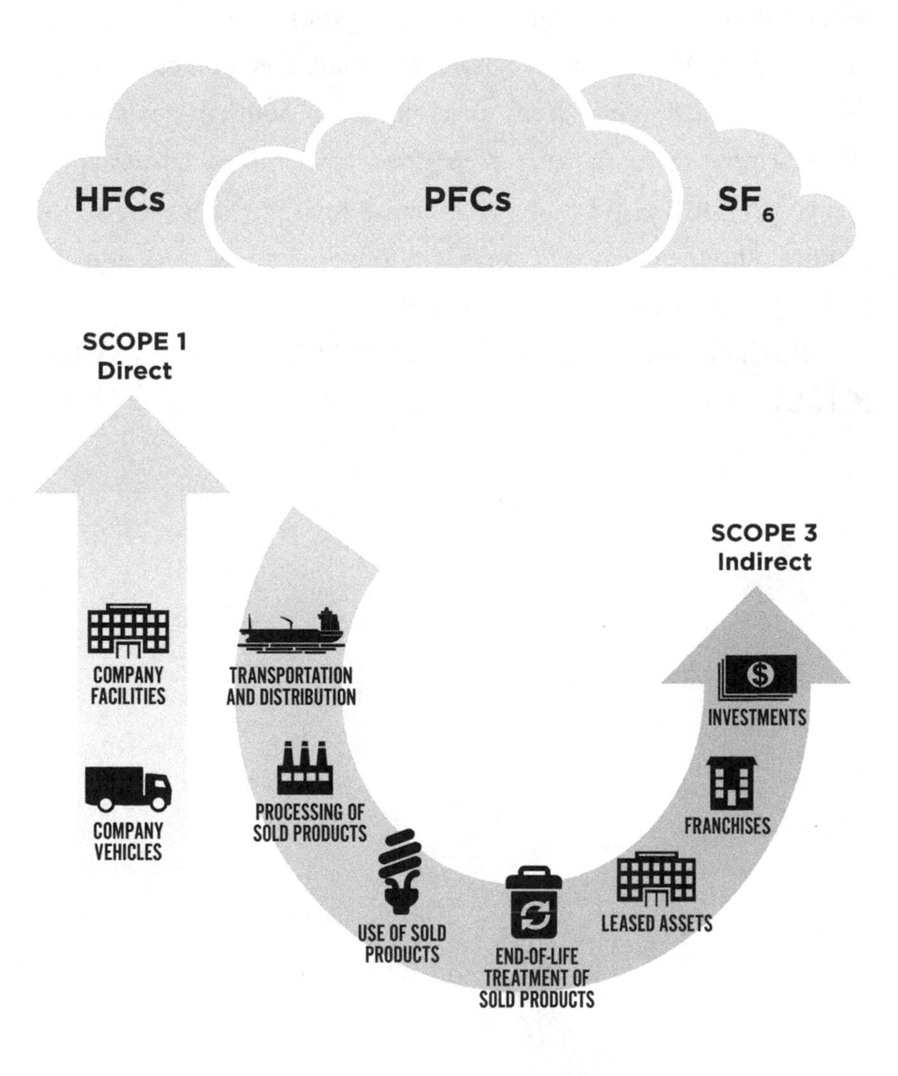

REPORTING COMPANY >>>> DOWNSTREAM ACTIVITIES >>>>

THE HGA HOUSE

With our three goals set and my investigations into methodology underway, the HGA sent out a letter to the architectural community. We received a response from an architect in Southampton whose client wanted to reconstruct their house after a fire had destroyed a large part of the structure, requiring a total gut renovation. The architect and the client were on board to work with us to design and build a house that would meet these three goals.

This house became known as the HGA House—and this is where ICEMAN began.

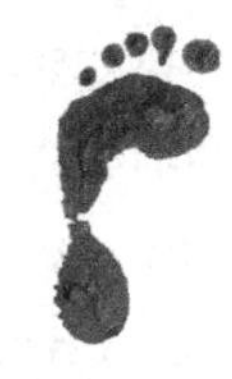

CHAPTER 3

FROM HGA TO ICEMAN

As I've said, there were no established metrics or protocols to measure and calculate the embodied greenhouse gas emissions of the construction stage of a building at the time the commitment to build the HGA House was made. We developed a methodology based on accepted protocols, including the WRI Protocol, carbon accounting of businesses, LCA, embodied carbon accounting of materials, offset mitigation, and offset projects.

Embodied greenhouse gas emissions of the construction stage come from the processing of raw materials, manufacturing of products, transportation of materials, and products in the supply chain and distribution system until it arrives at the jobsite. It includes labor and construction activities required for assembling and erecting buildings and how demolition and waste are handled. When construction is complete, the operations stage begins. The construction stage emissions are measured and calculated based on currently available databases and carbon accounting protocols.

We also needed to include all the energy used by all the subcontractors at each of their individual businesses. So for the construction of the HGA House to be carbon neutral, all the subcontractors had to agree to become carbon neutral themselves. We found subcontractors willing to commit to carbon neutrality.

The subcontractors who were members of the HGA committed to certify their businesses as carbon neutral. They were eager to lead by example and walk the talk by turning their businesses green. For our non-HGA subcontractors, we accounted for the Scope 1, 2, and 3 greenhouse gas emissions associated with this project, based on the percentage of their contract in relation to their gross revenues. For example, if a subcontractor's contract on the HGA House was $10,000 and their gross revenue for the year was $100,000, then we attributed 10 percent of their annual greenhouse gas emissions to the HGA House.

This included the embodied GHG emissions of materials used, transportation to the jobsite, energy such as electricity and heat used in construction activities, subcontractor Scope 1, 2, and 3 greenhouse gas emissions attributed to this project, demolition, waste disposal, and recycling. Since labor was subcontracted, the measurement also included labor in the construction of the house.

A FULLY INTEGRATED APPROACH

This level of coordination and cooperation required between all the subcontractors before we even got started building meant that we couldn't follow the traditional approach to the design and construction process. Ordinarily, construction follows a "design-bid-build" process. First, the design team—generally an architectural firm—designs a building and defines it by drawings and specifications. Some members of the design team may have construction experience but generally are not experts in construction. The design team generally creates the construction documents in a vacuum, with little input from construction experts.

The construction documents are then sent to several general contractors or construction managers who review the construction documents, interpret the content, and provide the cost estimate associated with construction—known as a bid. A general contractor or construction manager is selected and awarded the bid, and they begin construction. This process is repeated for all the subcontractors who bid their work to the general contractor/construction manager.

This fragmented approach often causes miscommunication, misinterpretation, and confusion concerning the intent of the design team. When integrating new green technologies, communication and collaboration needs to dramatically increase in order to achieve a successful outcome. Everybody needs to be on the same team, from the outset.

This is why we followed a concept of integrated design developed by the American Institute of Architects known as integrated project delivery (IPD). IPD brings the design team, owner, contractor, and subcontractors together during the early stages of the design process to foster collaboration, teamwork, information sharing, shared risks, and shared rewards. Contributions early on in the design process by construction experts in each field allow for more efficient integration of new technologies—some of which require input from several areas of expertise at the same time. When all the parties are in one room, information can be managed more efficiently and effectively. In fact, the LEED for Homes rating system awards points for implementing elements of IPD, because it is so necessary for successful green building.

It wasn't just the team that had to be fully integrated. To meet the goals we set, there had to be intense and complex interdependence between various building systems. The HGA developed the term "systems integrated home" to apply to the integration of multiple means and methods of design and construction to achieve maximum energy efficiencies.

TO MEET THE GOALS WE SET, THERE HAD TO BE INTENSE AND COMPLEX INTERDEPENDENCE BETWEEN VARIOUS BUILDING SYSTEMS.

Meeting the goal of net-zero energy is not as simple as slapping photovoltaic panels on the roof. The team studied reducing electrical consumption on a systems basis. We designed the heating and cooling of the house to be performed by a water source geothermal heat pump. Of course, the geothermal heat pump uses a significant amount of electricity. So the first course of action was to take the load off heating and cooling by super insulating the building envelope, installing energy-efficient windows and doors, and eliminating air infiltration. We also installed a variable speed well pump to help reduce the amount of electricity it takes to pump the water through the geothermal system.

We also installed an evacuated tube solar thermal system to generate energy for domestic hot water production. Our theory was that dumping excess hot water into the hot water coils of the air handlers in the winter would further reduce the load on the geothermal system.

We looked at the lighting system as well and found that LED lighting produces the same light output while reducing electric consumption by 85 percent compared to incandescent lights.[9] So we installed fifty-two LED recessed lights. We also installed a home energy monitoring system along with other smart home technology to monitor energy usage and automate several of the systems.

When it came to the products we used, since there is little LCA data on the manufactured products used in construction, our focus was on the embodied greenhouse gas emissions of the materials incorporated into the products and materials we used directly in construction. For example, there is no LCA data of a GE refrigerator; however, you can make an estimate of the greenhouse gas emissions of the materials used to make a GE refrigerator.

Some data on the embodied greenhouse gas emissions of materials does exist. While the embodied greenhouse emissions of a house cannot be completely, accurately measured at this time, we were able to make estimates of greenhouse gas emissions. As LCA data becomes more standardized and readily available, accuracy will improve.

---

9 CREE LED Lighting, "The CREE Difference," August 10, 2011, http://www.creeled-lighting.com/The-Cree-Difference/Efficiency.aspx.

OFFSETS AND CREDITS

As an organization, we purchased a carbon offset for whatever emissions we still had left. The offset we purchased was to help a family farm in Georgia that collected the manure from cattle and put it into an anaerobic digester. Anaerobic digestion is a process that converts biomaterial into biogas, which is very similar to the fuel we call natural gas—but instead of coming from fossil fuels deep within the ground, it comes from recycling biomaterial and is therefore renewable.

Biogas can be used to run generators, cars, machinery—anything you can run on natural gas, you can run on biogas. The farm uses the biogas they create to run a generator that produces electricity, which they use to power the farm. Whatever excess electricity they generate beyond what they consume, they put on the electrical grid.

We also earned carbon offsets for the greenhouse gas mitigation we performed as a result of landfill avoidance as part of the EPA's Waste Reduction Model (WARM), a protocol for tracking and reporting greenhouse gas emissions from solid waste.

Any kind of biomaterial—from food scraps to animal waste to wood products—when put into a landfill, decomposes and creates methane, which as we have said is thirty-four times more potent than CO_2 in causing climate change. This is why it's so important to prevent biomass from going into landfills. So the EPA created an incentive to keep biomass out of landfills. The EPA maintains an online calculation of carbon offsets in WARM. If you reduce the amount of waste you have going to landfills, you get a carbon credit from the EPA that goes toward your carbon-neutral status.

The HGA House had a good deal of waste, since the fire had damaged much of the structure and required a significant amount of deconstruction. Because we recycled a certain amount of wood material, we got a credit from the EPA through WARM for 107 metric tons of CO_2, which we could then take off of our carbon footprint.

The credit from the EPA worked hand in glove with our LEED certification, which also required us to recycle our biomass. LEED doesn't measure your carbon footprint. Instead, they say you are going to earn X number of points for this and X number of points for that. So we earned a certain number of points for recycling our wood waste products. The LEED points we earned for recycling waste materials provided a surprise benefit by also providing offsets, which mitigated the cost of purchasing carbon offsets.

FULLY CERTIFIED

Throughout the process, we worked with experts at the top of their fields. We retained the firm Verus Carbon Neutral to advise us, assist in measuring and calculating greenhouse gas emissions, and certify the house to be carbon neutral in the construction phase. That firm was run by the son-in-law of Dr. Richard Sandor. Dr. Sandor is known as the Father of Carbon Trading, a foundational concept in cap and trade. He founded both the European Climate Exchange in England and the Chicago Climate Exchange, which was eventually bought by the Intercontinental Exchange. His son-in-law's company specialized in measuring the carbon footprints of businesses; Dr. Sandor actually sat on the board of the company.

To confirm our carbon-neutral status, Verus performed a carbon audit to determine the greenhouse emissions associated with the construction stage of the building and factor in the offsets we had through carbon mitigation programs and the certified carbon offsets we purchased.

Verus also performed Scope 1, 2, and 3 carbon audits on all the

HGA member subcontractors, and the members purchased carbon offsets from Verus to offset 100 percent of their greenhouse gas emissions. Each member was certified carbon neutral.

certifies that

HAMPTONS LUXURY HOMES
AND ITS SUBSIDIARIES

have offset 100% of their
Carbon Dioxide Emissions.

Verus Carbon Neutral has retired **156** metric tons of CO_2e Offsets from the Chicago Climate Exchange on behalf of Hamptons Luxury Homes and its subsidiaries.

Certified by ____
Expires October 15 2010
Offset Project # CH4-WFF-1666

Applying all this science and all these processes and protocols worked. And not only did it work, but it was also confirmed by the third parties we had hired. Throughout the process, we had to have these things certified as we went. It wasn't just Verus; we also had a LEED rater who helped us through the LEED certification process. We had to justify and explain what we were doing to all these people, so we had checks and balances to certify every step of our process.

Our house was completely carbon neutral. We were the first builder in the US to be certified carbon neutral, as were our major subcontractors—HVAC, plumbing, electrical, etc. We also succeeded in our LEED goal: the total number of LEED certification points the HGA House earned was 104; the minimum number of points needed to attain LEED for Homes Platinum Certification is 100—so the HGA House was certified LEED for Homes Platinum.

I used the common carbon metric to measure the reduction of the carbon footprint in terms of the energy the building would use in the operations phase. We replaced all the energy that would have come from fossil fuels being used for electricity and heating with renewable energy and measured that over the life span of the building. This saved 70 percent of the energy that was being used in this house.

The organization Residential Energy Services Network developed a Home Energy Rating System (HERS) to measure the energy efficiency of a home through a series of tests and inputs, based off a standard new home built according to the 2006 International Energy Conservation Code. Obtaining a HERS Index is a prerequisite for the LEED for Homes rating system.

To claim a building is net-zero energy, the building needs to operate for one year while energy usage is monitored. If the building produces equal to or more energy than it consumes, the building is net-zero energy. While we did not reach this goal, the HGA House earned a HERS Index 25, which translates to a projected 75 percent energy reduction from the 2006 Standard New Home.

The HGA House was a success, especially when it came to the construction phase. This was hugely important—and not just for the HGA.

A BLUEPRINT FOR A REVOLUTIONARY CONCEPT

The greatest potential for low-hanging fruit in cost-effective, quick, deep greenhouse gas reduction and mitigation is found in the construction industry. Buildings worldwide account for a combined use of 40 percent of global energy and are responsible for one-third of the world's greenhouse gas emissions.[10] Buildings account for such a large amount of greenhouse gas emissions, so anything we do to reduce those emissions throughout the industry has a huge global impact on the carbon footprint of the whole world.

Our goal was to measure the greenhouse gas emissions embodied in the construction stage of the HGA House, and we developed a methodology to do that. That methodology is now the beginning of a framework to measure embodied greenhouse emissions in the construction of all buildings. The HGA House proved what the United Nations Environment Programme Sustainable Buildings and Climate Initiative found in their *Buildings and Climate Change: Summary for Decision-Makers* report in 2009: that with currently available and proven technologies, the reductions in energy consumption possible on both new and existing buildings are estimated to achieve 30–80 percent.[11] And it's not just a win for the environment: when the costs of implementing the energy reduction technologies are offset by energy savings, there is potential for a net profit over the life span of the building.

But it doesn't stop there. The work we did on the HGA House has implications far beyond the construction industry. Working through the HGA House, measuring the carbon footprint of the construction stage, applying these processes and protocols, successfully creating a carbon-neutral house, I realized: we're onto something here. Construction is essentially the same as manufacturing. So if something works in the construction of a house, it will work in the manufactur-

10 UNEP Sustainable Buildings and Climate Initiative, *Buildings and Climate Change: A Summary for Decision-Makers*, UNEP DTIE Sustainable Consumption and Production Branch, UNEP, 2009.

11 UNEP Sustainable Buildings and Climate Initiative, World Resources Institute, *Common Carbon Metric for Measuring Energy Use and Reporting Greenhouse Gas Emissions from Building Operations*, UNEP, 2010.

ing of any product. We could do this same thing to every product that is manufactured. And if we could do this for every product manufactured, if we could reduce the carbon footprint of manufacturing, from the raw materials through the finished product, it would have a huge impact on our environment.

IF WE COULD REDUCE THE CARBON FOOTPRINT OF MANUFACTURING, FROM THE RAW MATERIALS THROUGH THE FINISHED PRODUCT, IT WOULD HAVE A HUGE IMPACT ON OUR ENVIRONMENT.

When I realized that, I realized I was onto something pretty substantial—even revolutionary. So I sat down and started writing what I conceived: mathematically converting carbon neutrality data into an objective indexing system, easy for consumers to understand. I say I conceived this concept, but really, I *perceived* it. I saw that what we did with the HGA House could be applied to anything. And when I perceived ICEMAN, I perceived it in its entirety, beginning to end.

When I perceived it, I sat down at my home computer to write up an executive summary of this idea. Nobody had put together a methodology like this before. Measuring the carbon footprint of the construction stage was a new concept. I had to invent and define new terms. I started writing Saturday morning and didn't stop writing until it was finished on Sunday evening. I barely left my desk; I didn't sleep. On Monday morning, I sent it to my corporate attorney.

My attorney was blown away. As soon as he had read it, he called me. "Frank," he said, "This is really important. We have to protect it."

He referred me to an intellectual property attorney firm in Washington, DC, who filed the application for the service marks, for ICEMAN as well as all the new terms I invented in my executive summary. Service marks are a way to protect a concept. When a product is invented, a patent can be applied for. But a methodology is not an object. You can't draw a picture of it and get a patent. So, to protect a methodology, an idea, a concept, one needs to get a service mark. We filed these service marks with the US Patent and Trademark Office as well as with the European Union and Norway.

ICEMAN went through the approval process to get the service marks very quickly. As part of the process, the idea has to be used for a period of time to prove the marketability of the idea, which I did. After that, the service marks were determined to be indisputable. It's a completely original concept.

After the service marks were certified, I met a scientist at an event known as Green Drinks, who had degrees in applied physics, nuclear engineering, and mathematics. I asked him out for lunch and sent him a copy of the executive summary of ICEMAN in advance of our lunch.

We had barely sat down at the table before the scientist gave his assessment: ICEMAN was as significant a concept as the bar code.

"Really?" I asked in disbelief.

"Yes," he said. "This is very important."

This scientist introduced me to several research scientists and professors at Stony Brook University. The Stony Brook scientists took me under their wing and gave me credibility in academia. I was invited to speak at the Advanced Energy Conference in New York City, hosted by the Advanced Energy Research and Technology Center at Stony Brook.

In the audience was the editor of the American Institute of Physics *Journal of Renewable and Sustainable Energy*. He sent me an invitation to submit a scholarly paper on the HGA House, which included my application of the WRI Greenhouse Gas Protocol to measure the carbon footprint of the construction process. When the paper was peer-reviewed and published, I received an invitation to join the Society of Physical Scientists, as well as invitations to speak around the world. Scientists, experts, consultants—all at the top of their field in their industries—reviewed my process and findings and agreed with me.

I've shared how the real-world application of science and established protocols birthed the ICEMAN concept. Now, let's look at how ICEMAN works.

CHAPTER 4

THE MECHANICS OF ICEMAN

ICEMAN—International Carbon Equivalent Mechanism Attributed to Neutrality—is a methodology that applies well-accepted sciences developed for the calculation of greenhouse gas emissions to provide a quantitative measure of factors that reduce or mitigate greenhouse gas emissions. These quantitative measures are converted into an indexing system that represents a percentage of carbon neutral. The ultimate goal is for all products to fully offset their embodied greenhouse gas emissions. Not all products will be able to achieve full carbon neutrality. That's why a measurement system like ICEMAN's, based on a percentage to carbon neutral, will be most useful for manufacturers and consumers alike.

An index number of fifty indicates the product, process, or service is 50 percent carbon neutral; an index number of one hundred indicates it is 100 percent carbon neutral. The index number mathematically defines the greenness of a product based on the product's embodied greenhouse gas emissions.

The index number will be certified by ICEMAN and registered on a website where the public can access the information. The index number will also be incorporated into a logo certification mark that will be licensed to businesses, which they can place on packaging, websites, advertising, or any means the business uses to promote itself or its products or services. A consumer of those products or services can then use that attribute to evaluate those products or services, along with attributes like price, quality, and availability. This enables the market forces of competitive advantage to be a contributor to the overall reduction of greenhouse gas emissions.

ICEMAN is made up of several different components, which add up to the overall ICEMAN mechanism.

THE ICEMAN CARBON FACTOR

The ICEMAN Carbon Factor is the measurement of overall greenhouse gas emissions created by products, materials, transportation, business operations, and manufacturing operations used in the production of a product. The ICEMAN mechanism calculates and adds up the carbon footprint of a product throughout the supply chain, for every component and material—every time it is transported, every time a business operation touches it, all the way back to when the natural resources used to create its components are taken out of the ground.

When we look at the wood in our windows, we don't think about the impact it has. But if you go all the way back in the supply chain and look at everything involved in the process from when the log was taken out of the woods until the window was installed in your house, you can see the full carbon footprint of that item.

The mechanism begins accounting for greenhouse gas emissions from the operation that takes the natural resource out of the ground.

It includes emissions created by the transportation of the natural resource to the next part of the manufacturing process, emissions created by the process of converting the natural resource into a usable material or subcomponent, emissions created by the transportation of that material or subcomponent to the next part of the process, emissions created by the next part of the manufacturing process, and on through until the final product is finished and on the shelf.

The Carbon Factor is based on accepted sciences and standards developed as a result of established climate change protocols. Since the world began focusing on greenhouse gas emissions, scientists and organizations—both governmental and nongovernmental—have developed complex sciences and protocols to establish and standardize corporate accounting and mathematical calculations of greenhouse gas emissions. One example is WRI's Greenhouse Gas Protocol we used in building the HGA House. WRI is composed of a diverse group of government and nongovernment organizations, from the UN to the US Green Building Council.

Standards such as this will be adopted and integrated into ICEMAN, since ICEMAN is the application of this science. Any time the protocols change, we adapt ICEMAN to match the most up-to-date scientific information.

These aren't new concepts. These aren't new processes or protocols. These structures are already established. There are already well-established mechanisms for accounting and reporting greenhouse gas emissions. In fact, it is neither very difficult nor very expensive to get certified as carbon neutral. There are many organizations around the country that specialize in certifying carbon neutrality. You can even certify yourself as an individual! There are websites where you can enter your information and answer questions to calculate your carbon footprint; you'll then be given the opportunity to purchase offsets to cancel out your carbon footprint and establish yourself as carbon neutral.

There are businesses that specialize in these mechanisms, providing auditing and consulting services to organizations, assisting them in reporting and mitigating greenhouse gas emissions. As we've discussed, this is what we did for the HGA House. We had a third party certify and provide offsets for us to purchase to certify ourselves carbon neutral.

When we certified the HGA House in the construction stage carbon neutral, we formed a partnership with Verus Carbon Neutral to provide business auditing services. Verus had a special relationship with the Chicago Climate Exchange and Dr. Richard Sandor, who serves on their Board of Advisors. There is a high level of integrity and conformity to the protocols and mechanisms accepted by the Chicago Climate Exchange to calculate and measure the greenhouse gas emissions of business operations. The partnership with Verus ensures that the calculations integrated into the ICEMAN mechanism maintain the well-established and accepted standards for the accounting of greenhouse gas emissions. Since the completion of the HGA House, the Chicago Climate Exchange, Chicago Climate Futures Exchange, and European Climate Exchange were purchased by the InterContinental Exchange (ICE) in 2010.[12]

THE ICEMAN CARBON FACTOR OFFSET

The Carbon Factor will also take into account actions that companies are already taking that may impact overall carbon emission calcula-

12 Joel Kirkland, "Sale of Chicago Climate Exchange to ICE Reinforces Weak Carbon Market," *The New York Times*, May 3, 2010, https://archive.nytimes.com/www.nytimes.com/cwire/2010/05/03/03climatewire-sale-of-chicago-climate-exchange-to-ice-reinfo-362.html.

tions, such as sustainability, renewable energy usage, carbon offsets, the acquisition of carbon credits, and any carbon emissions mitigation or reduction programs, such as carbon sinks, clean development mechanisms, or joint implementation projects. These carbon offsets will be approved, certified, and tracked separately by the Carbon Factor Offset to maintain credibility and a high standard in accepting carbon offsets for the reduction of a product's carbon footprint.

The Carbon Factor Offset is a measurement of the carbon credits required to offset the Carbon Factor. The Carbon Factor Offset value is tracked alongside the Carbon Factor value. The total carbon offset values of manufactured products are the pro-rata of the offsets associated with the raw materials, components, subproducts, or services that make up the product. At any point along the supply chain, carbon emissions may be offset in part or in whole to become carbon neutral or partially carbon neutral.

The calculation of the reduction of carbon emissions associated with reduction programs or mitigation will follow well-established and standardized protocols. The reduction will be subtracted from the overall carbon emissions value, lowering the product or service's Carbon Factor. This formula is visualized here, showing the greenhouse gasses involved in creating a single window.

An example of this was used in certifying the construction stage of the HGA House carbon neutral. Recycling the wood waste was used to gain points in the USGBC LEED for Homes Platinum certification. There was a surprise benefit, described more fully in chapter 3: the EPA WARM calculates and provides carbon offsets for landfill avoidance that can be used to offset the entire carbon footprint. We were able to use the 107 metric tons of carbon emissions as an offset, increasing the carbon credits to reduce the overall carbon footprint.

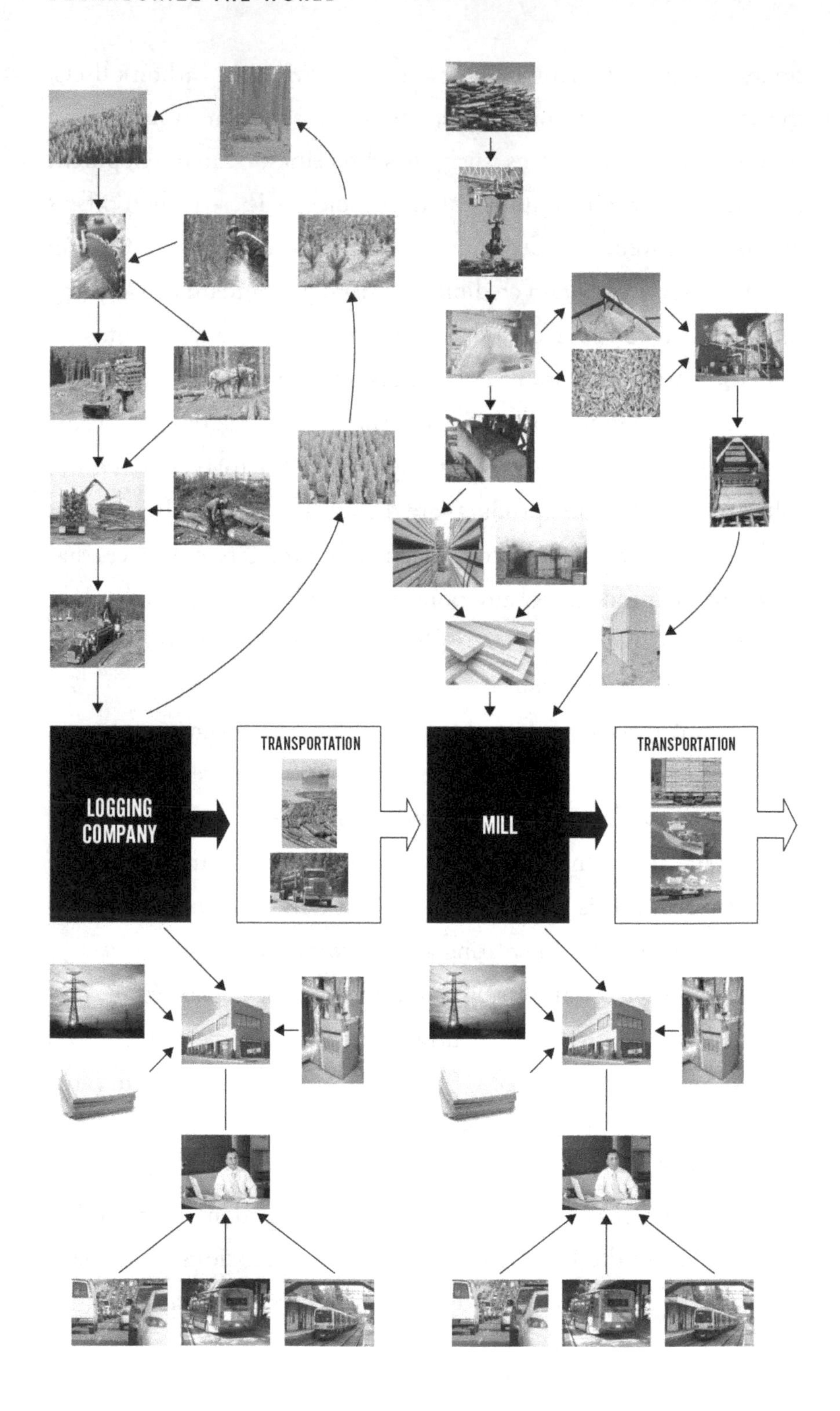
TRANSPORTATION
LOGGING COMPANY
MILL
TRANSPORTATION

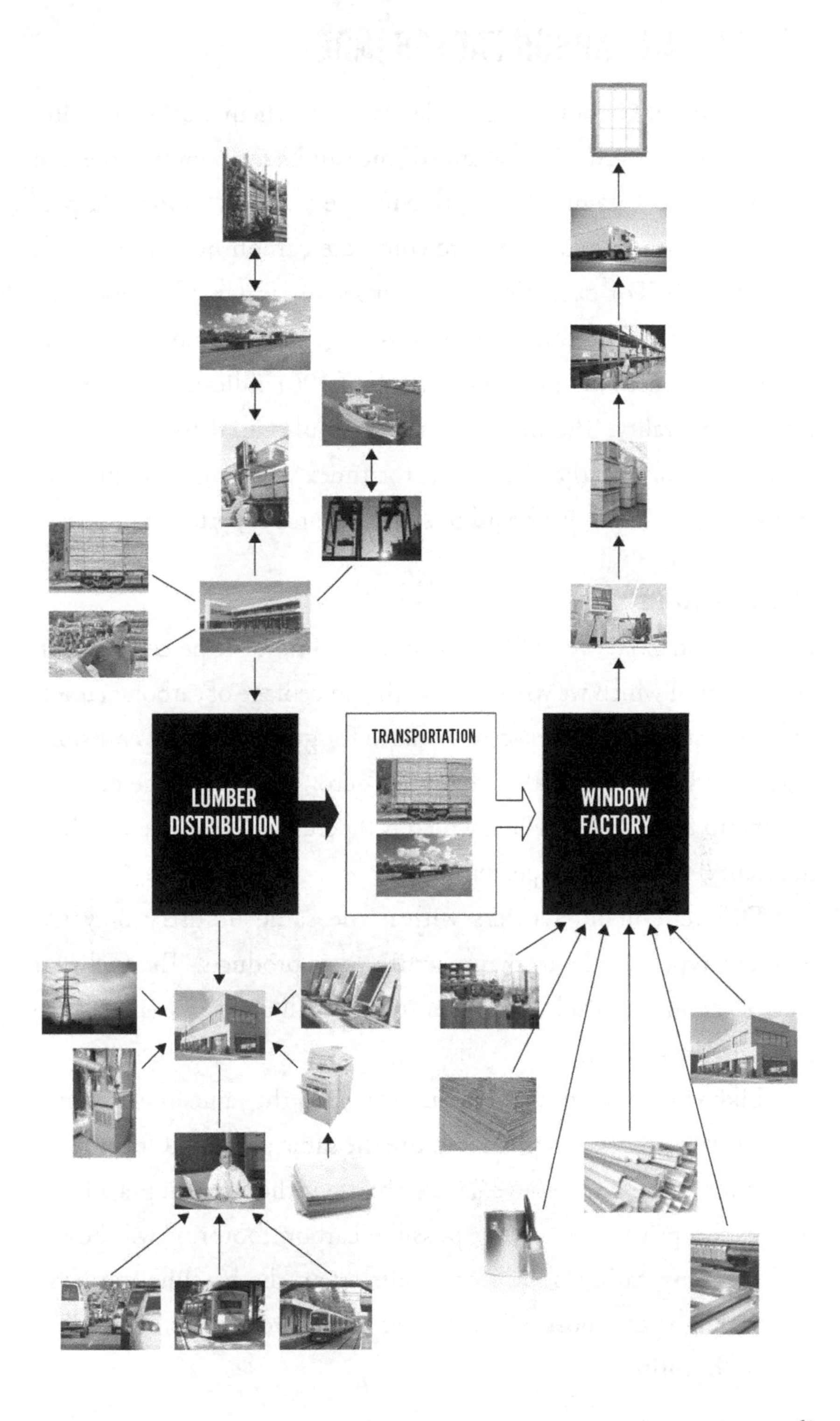
TRANSPORTATION
LUMBER DISTRIBUTION
WINDOW FACTORY

THE ICEMAN CARBON FACTOR INDEX

The ICEMAN Carbon Factor Index is the mathematical conversion of the Carbon Factor into a unit of measurement between one and one hundred. This number represents the Carbon Factor as a percentage of carbon neutral, where complete carbon neutrality equals one hundred. For example, a Carbon Factor Index of eighty (CFI 80) indicates the product or service is 80 percent carbon neutral. A Carbon Factor Index of one hundred (CFI 100) indicates 100 percent carbon neutrality. The mathematical formulas used to convert the Carbon Factor into the Carbon Factor Index will remain proprietary and will be verified by a board of scientists and experts.

The Baseline

In every industry, we will establish a baseline of 0 percent carbon neutral, from which we will calculate the percentage of carbon neutrality. The result of the worst-case scenario for greenhouse gas emissions in each industry will be the baseline of zero. The percentage number counts up from that baseline, which is 0 percent, to complete carbon neutrality, which is 100 percent.

Different manufacturers within the same industry may use different types of fuel to manufacture their products. The fuel that creates the most greenhouse gas emissions will be used as a factor to calculate the baseline.

Likewise, different manufacturers within the same industry may use different processes to manufacture the same product. One process may emit more greenhouse gasses than another, creating a higher carbon footprint. The highest possible carbon footprint will be the baseline. Any carbon footprint smaller than the baseline footprint will highlight the most efficient method in producing the smallest carbon footprint.

The difference between dry kiln and wet kiln methods of producing clinkers for concrete is a great illustration. The wet kiln process is less energy efficient than the dry kiln process. Therefore, concrete from two different manufacturers who use these two different methods will have different carbon footprints.[13,14]

Different manufacturers producing the same product within the same industry may also have different carbon footprints based on what electrical grid they are on. The combination of electric generation types within a grid determines the calculation of greenhouse gas emissions per kilowatt hour. The baseline for the industry will include the highest emitting type of grid in calculating the worst-case scenario.

The worst-case scenario for an electrical grid in the US is a coal-fired electric grid. That is the type of grid that creates the greatest amount of greenhouse gas emissions per kilowatt hour of electricity produced. That is the baseline, the 0 percent Carbon Factor Index value. It all goes up from there, all the way up to renewable energy, such as solar, wind, and hydroelectric generation, which produces no greenhouse gas emissions.

We can use that baseline for electricity grids across every industry. The more renewable energy a grid uses, the closer it is to being carbon neutral. For example, the Niagara electric grid is 100 percent renewable hydroelectric energy. The grid we have here on

THE MORE RENEWABLE ENERGY A GRID USES, THE CLOSER IT IS TO BEING CARBON NEUTRAL.

13 Engineering Discoveries, "Difference between Wet and Dry Process of Cement," November 6, 2019, https://engineeringdiscoveries.com/difference-between-wet-and-dry-process-of-cement/.

14 Lisa J. Hanle, Kamala R. Jayaraman, and Joshua S. Smith, "CO_2 Emissions Profile of the US Cement Industry," 2004, https://www3.epa.gov/ttnchie1/conference/ei13/ghg/hanle.pdf.

Long Island in New York uses some renewable energy, some number two fuel oil, and natural gas energy.

So a manufacturer who uses electricity from a grid that distributes mostly coal fire–generated electricity will have a higher carbon footprint than a manufacturer who uses electricity from a grid that distributes more electricity from renewable sources. If a corporation uses electricity from a renewable energy grid, it already has an advantage. It will already be closer to carbon neutral.

In a situation where a manufacturer installs renewable energy such as solar panels or a solar farm to produce all their electrical consumption needs, the manufacturer will have an advantage, since the electricity used to manufacture the product has no carbon footprint.

The type of grid a company uses is combined with other factors to create the baseline for that product. To go back to our concrete clinker illustration, manufacturing clinkers in a coal-powered electric grid will yield a larger carbon footprint than manufacturing clinkers in a hydroelectric grid. So, in a simplified example, a manufacturer using the wet kiln method on a coal-powered electric grid would be the worst-case scenario—the baseline, 0 percent—for manufacturing clinkers.

It is important to understand and reiterate that all carbon measurements are cradle to gate. In the beginning we will be using broad strokes in measuring the carbon footprint, using Scope 1 and Scope 2 emissions. As we ramp up, we can get more granular and begin to include Scope 3 upstream activities. The idea is to pick the lowest-hanging fruit without getting bogged down in the weeds.

We understand, when installing solar panels, that the solar panels themselves have a carbon footprint. We will account for the carbon footprint of the solar panels during their manufacturing process, not in the production of electricity. We believe we will capture the majority of carbon emissions with this method. If in the future it becomes obvious we aren't as successful in reducing global carbon emissions as we had anticipated, we will always be able to become more granular, possibly including Scope 3 downstream activities. It is much more important to get started as soon as possible, with as much of an impact in reducing carbon emissions as possible.

At all levels of granularity, the baseline will be established according to the most up-to-date science, coordinating with both government and nongovernment organizations like the WRI, academic institutions, and the EPA and DOE.

Calculations

Once the baseline is established, we take the greenhouse gas emissions data from the individual company or manufacturer. Using that data and companies' own cost accounting, we calculate the emissions of the specific product that is produced. The carbon-neutral status is calculated based on the greenhouse gas emissions emitted during the manufacturing process, plus the combined embodied carbon footprint of all the raw materials and components of the product.

The pro-rata sum of the carbon emissions of all the parts—the raw materials, subproducts, services, and manufacturer operations—determines the index of the manufactured product. Once the Carbon Factor Index is known for raw materials, components, subproducts, services, and manufacturer operations—having gone through this process with all the companies that process and produce those materials, components, subproducts, etc.—the total Carbon Factor Index can be calculated for every manufactured product using a simple mathematical calculation of the pro-rata sum of the parts of the product. For example, if a raw material represents 10 percent of that product, you prorate the carbon footprint of that material, just as you would prorate the cost of the raw material.

Having already calculated the Carbon Factor Index for all the components and raw materials that make up the product, we can look up the index numbers for those component and raw materials, and from that, using a mathematical calculation, determine the Carbon Index Factor of the finished product. Knowing the Carbon Factor Index of all components, materials, and operations involved in the creation of a product will eliminate the mystery of how green that product is.

The calculation of the pro-rata sum of the parts may differ, since different units of measurement are used from one industry to another. For instance, the unit of measurement for volume in the lumber industry is board-foot. The unit of measurement for volume in another industry for concrete may be cubic yard or cubic meter. Some industries use surface area measurements, like square feet or square meters. In some industries, units of measurement may be by weight.

Some industries may measure the ingredients used to manufacture the product, so a percentage of each ingredient needs to be calculated in determining the pro-rata sum of the parts. Some industries may use a percentage of completion based on cost, so cost or percentage of cost needs to be used to calculate the pro-rata sum of the parts.

However, every industry has a standardized cost-accounting method. Cost accounting is the process of identifying the detailed and aggregated costs of producing a product or service. Each industry has its own cost-accounting methods, and ICEMAN uses those same cost-accounting methods in accounting for carbon emissions, incorporating the industry's method to calculate the pro-rata sum of the parts. In each industry, this standardized method will be used to calculate the Carbon Factor Index. Therefore, the Carbon Factor Index calculations may vary from one industry to another, but the mathematical formulas will be standardized within each industry to maintain continuity.

Units of measurement also differ from country to country. Some countries use the imperial measurement system, while others use the metric system. It is a simple conversion to go from the imperial system to the metric system. For example, the international measurement for greenhouse gasses is metric ton. The simple conversion of metric ton equals 1.10231 imperial tons.

Regardless of the unit of measurement and cost-accounting method used, once the Carbon Factor Index is calculated, that value will be consistent across all industries and all countries. The Carbon Factor Index is a sound scientific and mathematical basis for calculating carbon neutrality in a simple and easy-to-understand value system. That system can be standardized across all industries, regardless of the units of measurement used in each industry.

In order to get the Carbon Factor Index value of a specific product, we will gather some additional information from the company. The company's overall emissions data will most likely already be reported to the EPA for the larger companies; the EPA will also need to collect overall emissions from the smaller companies. We will then work with manufacturers to get the information we need about individual products, to reconstruct the entire supply chain, from the raw material to the finished product on the shelf.

Some manufacturers manufacture several different products. They will have reported their carbon emissions as a whole; we will gather additional information to calculate the emissions of the specific product. We would break this down using the corporation's own cost-accounting structure for the specific products they produce. Corporations always track their costs for each individual product. After all, they need to know how much to sell the product for! We will use the same structure for carbon accounting.

The Current Scope

Currently, ICEMAN only covers Scope 1 and Scope 2, as defined by the WRI Greenhouse Gas Protocol. It does not factor in what happens to the product once it is in the consumer's hands, and what happens to the product at the end of its life cycle—the part of a product's life cycle known as "downstream."

There are some products where it is possible to easily calculate the downstream carbon footprint. I heat my house with recycled wood products—pellets made out of recycled wood waste products, furniture making, sawdust, etc. Since it is recycled material, it is considered renewable energy. It's not going into landfills and creating methane. However, burning wood does create carbon emissions. So the company I buy my pellets from has created their own carbon offset project. They account for the smoke going out of chimneys and have committed to planting a tree for every forty-pound bag of pellets burned. This is where downstream accounting actually works. They know you are going to burn their product, so they know what emissions it will produce.

However, that is not true for most products. There's no way to know whether an individual who purchases a plastic product will recycle that product—or, if they do, where they recycle it, as every township and municipality has different recycling rules and systems.

There is certainly a market for plastic. Companies will purchase recycled plastic from towns and reuse it for other products or even use it to create additional plastic. The towns, instead of paying to dispose of the plastic, can sell it to somebody who is going to use it. Plastics are a good example of downstream recycling. But you can't tell by looking at a company or a product whether someone is going to recycle that product or just toss it.

Once ICEMAN has been adopted, we can start addressing

downstream effects. When WRI developed their Greenhouse Gas Protocol, in the beginning they used broad strokes. As the protocol was developed and became accepted, they started getting into more detail, including more and more things to measure. Now, they've reached a point where it is very complex and very comprehensive. However, it took time to get there. If we tried to be completely comprehensive right off the bat, before implementation, it would become incredibly unwieldy.

As ICEMAN matures and becomes accepted and utilized, we can start incorporating more measurements. Just like the WRI Greenhouse Gas Protocol, we can add in more detail as we go. We can drill down and get more and more comprehensive in determining the carbon footprint of everything. We could even get to the point of measuring the carbon dioxide expelled from our own lungs! But we need to start with the low-hanging fruit. What is the carbon footprint from a broad-strokes viewpoint? So, in the beginning, we'll focus on Scope 1 and Scope 2, and in the future, we may bring in Scope 3.

Beyond Carbon Neutral

To reduce its carbon footprint and earn a higher Carbon Factor Index value, a company first does mitigation, meaning they make as many adjustments as possible in the entire cradle-to-gate manufacturing process. But if a company cannot achieve complete carbon neutrality through mitigation alone, they need not despair. For whatever greenhouse gas emissions a company can't mitigate, it can purchase offsets. Offsets are projects that are actively taking carbon emissions out of the atmosphere, like the offset we purchased for the HGA House: the money we paid for the offset went to the family farm, which helped the farm pay for the equipment to convert manure into biogas through anaerobic digestion. The biogas was then used to run a generator to produce electricity consumed at the farm.

A manufactured product may claim carbon neutrality by either making sure every component used in the manufacturing of the product has a CFI 100, or by purchasing carbon credits equivalent to the pro-rata sum of the component's Carbon Factor Offsets to make up the difference. Through mitigation, a company might achieve 70 percent carbon neutrality—a CFI 70. Then, they can go to the market and purchase carbon offsets for the remaining 30 percent in order to be certified 100 percent carbon neutral. This closes the circle between those that provide offset projects, which take carbon out of the atmosphere, and those that can't mitigate their carbon emissions but have determined they will have a greater competitive advantage with a CFI 100, and so have decided spending the additional money on carbon offsets is a prudent business model. Essentially, it is like spending more advertising dollars to gain more market share.

Theoretically, a manufactured product or business operation may have a Carbon Factor Index greater than one hundred, if carbon factor offsets are purchased that exceed the overall Carbon Factor Offset value of the product, or if the business purchases carbon credit offsets greater than its carbon emissions and the embodied component parts of the product. If the business participates in carbon capture or carbon sequestration projects or any other program that mitigates greenhouse gasses, the business may be able to offset more than the emissions they have created.

A Carbon Factor Index value higher than one hundred may become a marketing tool or a competitive advantage for the companies that supply raw materials, subproducts, and services to manufacturers, since their higher Carbon Factor Index value will have a positive impact on the Carbon Factor Index value of the product the manufacturer is making.

THE CFI CERTIFIED SERVICE MARK

Once adopted and implemented, the Carbon Factor Index value will be incorporated into a logo certification mark that will be licensed to businesses to place on their packaging, websites, advertising, or any means the business uses to promote themselves or their products or services. The certification mark will provide an easily understood verification of how green a product is. A certified label identifying the product's relationship to carbon neutral would be licensed to businesses.

The examples of the Carbon Factor Index label below have been designed with a black "CFI" for values below one hundred, indicating carbon emissions present in the manufacturing of the product. When the values reach one hundred or greater, the "CFI" is gray, indicating neutral—no carbon emissions present in the manufacturing of the product—or greater than carbon neutral.

The Carbon Factor Index value can be easily understood by consumers. When the CFI certified mark is put on a website or packaging or in an advertisement, consumers can easily make an evaluation of that product or service's carbon footprint and compare it to the carbon footprint of other products/services. A CFI of one hundred indicates that the production of the product did not have any adverse impact on the environment from carbon emissions; the lower the CFI value gets, the worse the manufacturing of that product is for the environment.

It couldn't be simpler: a product with a Carbon Factor Index of one hundred is greener than a similar product with a value of fifty.

The consumer will see the certified index value on the packaging, and it will give them and additional attribute by which to evaluate those products or services, in addition to attributes like price, quality, and availability. Just as a consumer can look at a food product and easily see if it has a certified organic label, a consumer could pick up a product, look at the ICEMAN label, and say, "This product has less of a carbon footprint than that one." If the product is more or less the same price and quality, a consumer will likely say, "I might as well buy the one that has less of an impact on the environment." And they will know that the product does, in objective fact, have less of an impact on the environment, because the CFI value is completely objective, based on a mathematical formula.

THE DATABASE

To verify certification of the Carbon Factor Index, the index will be listed on a website, in a searchable database open to the public. Manufacturers will need to stay current on their carbon accounting and reporting to remain certified. Carbon accounting is based on a one-year look back—the carbon emissions in the prior year. Each year that may change, and an opportunity is given to allow manufacturers to update their carbon emissions data. The database will also include Carbon Factor Offset values, since manufacturers may decide to reduce or offset all or part of their greenhouse gas emissions at any point along their entire supply chain.

Consumers and wholesale buyers of products will be able to look up and verify the Carbon Factor Index value of every product made.

Shoppers will be able to know the Carbon Factor Index value of any product before buying it.

The database website can also become a tool in and of itself. It can become a business hub that allows opportunities for companies to promote their low-carbon operations. The data will be organized by standard industry codes to facilitate effective target marketing by companies.

Say a manufacturer is producing widgets. Within the manufacturing corporation, there is a buyer who purchases the raw materials and components that make up the widget. That buyer is concerned with all the same attributes as a consumer—price, availability, quality, etc. Now, they can consider the carbon footprint attribute as well when considering which raw materials and components to purchase. The buyer can go onto the website and search for which companies are selling the materials and components with the highest CFI value. By purchasing materials and components with high CFI values, they can create a product with a higher CFI value.

When they first start out, many tech companies struggled with how to monetize their ideas. It took the various social media platforms a while to figure out how to target advertisements to users for revenue. ICEMAN, on the other hand, is a tool for people who are already ready to buy. There's no algorithm necessary to determine what the buyer wants or how to target them. When a buyer for a manufacturer searches for something in the database, it's because they are ready to buy it. That very moment, when the buyer is ready to pull the trigger to make a purchase, is the best opportunity to cross-market that buyer with a product that has a higher CFI value.

Therefore, manufacturers will want their product to have a high CFI value. When the buyer sorts the product for which they are searching by CFI value, manufacturers will want their product at the top of the list. It's direct marketing. Companies that have worked to reduce their carbon footprint will automatically be marketed to the buyer, who is ready to purchase. Successful marketing at this time will have greater results than marketing on any other social media platform.

REPORTING

In order to determine the Carbon Factor of a product, to set the base starting point, the "0" of the worst-case greenhouse gas emissions scenario for each product, to create the Carbon Factor Index, we need greenhouse gas emissions data from manufacturers. Where do we get this information?

This is where the government comes in.

Currently, the EPA already requires large corporations to report their carbon emissions. Many of these larger companies have already been reporting their carbon emissions, because they are multinational, and reporting is required in Europe under the cap-and-trade system. If a company sells their product in Europe, they are already reporting their carbon emissions to the governments of whatever countries they sell in.

Reporting is currently voluntary for smaller businesses. But small businesses make up over 99 percent of all business in the United States.[15] The EPA could easily expand the mandate to include small businesses, requiring them to report their emissions as well. It wouldn't be a significant cost even to a small business. After all, the smaller your business, the smaller your carbon footprint will already be!

15 JPMorgan Chase Institute, "Small Economic Activity," accessed May 2, 2021, https://www.jpmorganchase.com/institute/research/small-business/small-business-dashboard/economic-activity.

The EPA collects all this data and, with it, can assemble a database of the carbon emissions of every company in the United States.

Greenhouse gas emissions can be reported on a corporate federal tax return. Starting out, this reporting will be very simple, only covering Scopes 1 and 2 following the WRI Greenhouse Gas Protocol. This will amount to less than half a page of energy usage data, included in a company's corporate IRS tax return. As we implement ICEMAN, we will drill down and get more granular and comprehensive. But even then, a corporation should never need more than one page of reported data.

This is the only element of ICEMAN that requires any kind of government mandate. The government does not need to mandate that companies report under a certain amount of carbon emissions. Companies just need to report whatever their emissions are. The government does not need to mandate that any company or product maintain above a certain CFI value. In fact, the government does not need to mandate that companies obtain a CFI certification for any of their products. ICEMAN is completely voluntary.

Implementing ICEMAN doesn't include any regulation beyond requiring a company to report their greenhouse gas emissions. That is the only mandate we are asking for. We're not asking for a mandate saying, "You have to report this percentage or below. You have to achieve this CFI or above." We're not asking for a mandate requiring companies to reach a certain percentage by 2030. All we are asking is for companies to report their emissions, whatever they are.

If individual municipalities, towns, or even states want to mandate a certain CFI rating for particular industries, they can do so. This is what happened with the HERS rating: it is now written into the New York State building code that buildings must achieve a certain HERS Index. Some localities, like my town, have an even lower HERS Index requirement in their building codes. I was actually on the committee that advised our town board to lower the HERS Index, known as a stretch code. We wanted to build more energy-efficient homes—and people in our community were already doing it. But this mandate only affected buildings over 4,500 square feet, so it didn't have an impact on affordable housing or other low-income buildings.

Once the manufacturer reports its greenhouse gas emissions to the EPA, we will use that data to create an index number based on those reported emissions. Then, the market forces of competitive advantage will do the rest. It is that simple, that easy, and completely politically benign.

A PUBLIC-PRIVATE PARTNERSHIP

To implement ICEMAN quickly and efficiently, with integrity that will be accepted worldwide, the science community, government, and private businesses need to come together and contribute their expertise. Therefore, implementing ICEMAN will involve a public-private partnership structure with the United States government, and possibly governments in other countries as well. Most government organizations, including the EPA, have established structures for forming such a relationship.

In a public-private partnership with the EPA, the EPA would mandate emissions reporting from all businesses, as we talked about previously. The EPA would house all that information in a database and allow ICEMAN access to that database. ICEMAN would use that data to calculate the Index, and we would hold the CFI Index in a searchable public database.

This arrangement has automatic advantages for the EPA. Once the EPA begins collecting data from all businesses, not just large companies, the government would be better able to measure how the country is doing in relation to any goals the country has committed to as part of the Paris Agreement or any other agreement we might enter into. It is great information for the EPA to have. It can be used to make policy, to educate people, and to help them reduce their carbon footprints.

But a public-private partnership, which may also involve other government organizations like the DOE, will do more than just build the EPA database and give ICEMAN access to emissions reporting data. The work required to establish the baseline on which the Index is built, and to then set up the Index, is substantial. We need both personnel and funding to accomplish it. The government will need to allocate and fund the EPA for implementation. The DOE will be able to provide expertise, oversight, and grant funding to establish the baseline. A public-private partnership would allow the EPA and DOE to allocate both personnel and funding to complete the work to the highest standards with integrity.

With that funding, we can bring on academics and scientists from top institutions, as well as people from government organizations, nongovernmental organizations, and even the United Nations. We plan to invite distinguished scientists onto a Board of Science and Research, which will include the brightest minds committed to adopting the best sciences into the ICEMAN application. I've already started to bring academia into the project, discussing it with the research professors I'm working with on my green building initiatives. We may also put together an advisory board that includes corporate leaders, creating a kind of think tank to organize the whole structure.

Implementing ICEMAN is not a difficult undertaking; however, the scope is very large and will require expertise found in the private sector as well as the public sector. A private company will have greater success working across political party lines and international political boundaries and agendas. A private company will be able to more efficiently implement such a large undertaking.

AS ICEMAN EXPANDS, THE PUBLIC-PRIVATE PARTNERSHIP CAN EXPAND AS WELL.

As ICEMAN expands, the public-private partnership can expand as well. The government can be very useful in implementing a labeling system by requiring manufacturers to include the Carbon Factor Index for the customer's information, in a manner similar to organic food labels or labels that list ingredients or safety information. As we saw with the organic movement, any type of consumer labeling system usually requires government verification, oversight, or regulation once it reaches a certain level of use in the marketplace.

The Power of a Partnership

If you want a perfect illustration of the power of a public-private partnership between the government and a private organization, look no further than the COVID-19 vaccine. Generally, it takes years to develop a vaccine, test it, and get it approved by the FDA. Operation Warp Speed was announced on May 15, 2020. The US Department of Health and Human Services, Department of Defense, and several other federal agencies partnered with eight companies in the private sector, granting them between $38 million and $2.1 billion each to develop a vaccine. The FDA approved the Moderna vaccine for emergency use on December 18, 2020—a mere six months after the operation was announced—and the Johnson & Johnson vaccine on February 28, 2021. With the power of a public-private partnership, Moderna and Johnson & Johnson developed and brought two wildly effective and safe vaccines to market faster than has ever been done before.

You can accomplish amazing things with a public-private partnership working toward a common good. Although it may not seem like it on the surface, the climate crisis is as immediate a crisis as the COVID-19 pandemic. Through a public-private partnership, ICEMAN can help alleviate this crisis, just as Moderna and Johnson & Johnson have for the COVID-19 crisis.

But it's not just the partnership between ICEMAN and the government that will make it happen; it is also the power of the market forces of competitive advantage. The power of those forces cannot be overstated.

CHAPTER 5

THE POSITIVE FEEDBACK LOOP: HARNESSING MARKET FORCES

If the country is to start taking a more proactive approach to reducing greenhouse gasses, it needs to educate its citizens. While most of the industrialized nations of the world signed on and ratified the Kyoto Protocol, committing to reducing greenhouse gas emissions, the United States did not. As a result, most citizens of the US have little knowledge or understanding of the protocols, mechanisms, and science developed worldwide to reduce greenhouse gas emissions.

The Biden Administration is attempting to adopt an energy policy that will enable the US to join the rest of the world and make a commitment to reduce its carbon footprint. But the citizens of the US will need to catch up to the progress that has already occurred in other participating nations. US citizens may be faced with an exponential learning curve in the next few years.

ICEMAN will be able to assist in the learning curve.

Knowing the Carbon Factor Index of materials, products, business operations, and manufacturer operations will accurately demonstrate how green a product is. Since the average American isn't steeped in the science behind climate change and greenhouse gas emissions, placing these labels on a product will eliminate the mystery of whether or not a product is truly "green." Right now, nobody really knows the carbon footprint of any product or company. ICEMAN will reveal that information for everyone to see.

Knowledge is power. Once the consumer has access to that information, that will be the incentive for corporations to clean up their act. When food manufacturers were required to list nutritional information on packaging—and especially as label requirements were updated in recent years to be more accurate and convey information more clearly—it affected how people ate. According to a report in the *American Journal of Preventive Medicine*, consumers "reduced the intake of calories by 6.6 percent, total fat by 10.6 percent, and other generally unhealthy choices by 13 percent. They also increased vegetable intake by 13.5 percent."[16] In response, food companies on average reduced the amount of sodium in their products by 8 percent and the amount of harmful trans fat by 64 percent.[17]

Nutrition labels on food are effective because consumers generally have a basic level of knowledge about nutrition, about what is healthy or unhealthy. They know too many calories is unhealthy; they know too much fat or sodium is unhealthy. They know vitamins and protein are good. A consumer can look at a nutrition label and understand it, without any kind of degree in nutrition or food science.

This is what ICEMAN offers for carbon footprints: information

16 Dariush Mozaffarian and Siyi Shangguan, "Do Food and Menu Nutrition Labels Influence Consumer or Industry Behavior?" *STAT* (blog), February 19, 2019, https://www.statnews.com/2019/02/19/food-menu-nutrition-labels-influence-behavior/.

17 Ibid.

that any consumer, anywhere in the world, can understand, no matter their level of knowledge about environmental science, greenhouse gasses, or climate change.

A SIMPLE MEASUREMENT

Within the scientific community, the standard measurement for greenhouse gasses is metric tons. This is the unit of measurement generally used in the greenhouse gas protocols developed by scientists, government, and nongovernmental organizations.

But outside of the scientific community, this measurement is neither universal nor easily understood. We visualize greenhouse gasses as, well, gasses—floating invisibly in the air. As such, it is difficult to visualize the size, weight, or volume of a metric ton of greenhouse gas emissions. In fact, it's hard to comprehend the weight or volume of a gas in any metric measurement.

This is especially true in parts of the world, such as the United States, that use the imperial system of measurement rather than the metric. Consumers in the US may have difficulty conceptualizing how much a metric ton of greenhouse gasses is. The average consumer may know what a twenty-pound tank of propane for a barbeque grill looks like, but it is much more difficult to visualize what a metric ton of greenhouse gasses looks like. Without a conversion table in front of them, a US citizen may have difficulty understanding what a metric ton means.

If consumers cannot easily understand the measurements used to quantify greenhouse gasses, they will have a more difficult time understanding and embracing climate change science—and will therefore be less likely to embrace the cause of reducing greenhouse gas emissions.

To make the complex measurement of greenhouse gasses easily understandable to consumers everywhere, ICEMAN mathematically converts metric ton measurements of greenhouse gasses into a simple, universal indexing system based on a percentage value of carbon neutral. This index system can be universally understood by any consumer anywhere in the world, regardless of what measurement system they use or their level of knowledge about climate change science.

It will be very important to standardize the carbon indexing system, since products are manufactured all over the world. The label and index value need to be the same to all consumers in the world, regardless of where the product is manufactured.

THE CONSUMER-DRIVEN COMPETITIVE ADVANTAGE

After observing the marketing advantages of the present green movement, you can imagine how significant reductions of greenhouse gasses may be achieved through competitive advantage if "greenness" is verified through a mathematically based method. This is how ICEMAN enables the market forces of competitive advantage to be a contributor to the overall reduction of greenhouse gas emissions.

Consumers consider a number of attributes when they are deciding what product to buy. Price is an attribute. Quality is an attribute. Availability is an attribute. Now, carbon footprint can be an attribute as well. If there are two products that are more or less equivalent in terms of quality and price, a consumer may choose one over the other based on which has the smaller carbon footprint. It's just common sense that someone would want to pick the product that would have less of a negative impact on the environment.

With a CFI label, the consumer will be able to evaluate and

make a decision based on which product has the better Carbon Factor Index value and know that they have chosen the product that is, in fact, greener. As consumers continue to embrace green products, a company that can deliver accurate, understandable, and reliable information will have a competitive advantage in the marketplace. As such, manufacturers of products will gain that competitive advantage by advertising their Carbon Factor Index value and placing their CFI label on packaging.

Rather than any kind of government mandate or regulation, the consumer becomes the driver for the implementation of ICEMAN. Companies will recognize that consumers will choose products and services while taking into account the attribute of the CFI value. If a company does not include that attribute on a product, consumers will be less likely to consider it, and the company will lose market share. Because of this, I truly believe ICEMAN can be implemented completely voluntarily. We don't need to rely on government regulations or mandates; the market forces of competitive advantage will take care of implementation. The consumer is going to be educated. And when they are educated, when they can compare the carbon footprints of different products, they are going to choose the product that is better for the environment.

Brand Image

Moreover, corporations will have a powerful tool to help enhance their image by promoting the green attributes of their products or operations. Brand image is incredibly important. Companies are very concerned with what people think of them, especially in these days of rapid judgment and censuring on social media. Companies want to make sure they have a positive image. I'm sure we've all seen the commercials companies put out, touting how green the company is without talking about the product itself—or advertising the greenness of a product without considering the overall greenness of the company. The ICEMAN Index takes into account the carbon footprint of both the company and the product.

With people getting more and more onto the green bandwagon, being environmentally responsible is good for brand image. Everybody wants to be viewed as environmentally responsible. ICEMAN allows a company not just to *seem* environmentally responsible; it allows a company to demonstrate, as an objective, mathematical fact, exactly how environmentally responsible they are.

Being able to evaluate the greenhouse gas emissions of a product or service will influence consumers' purchasing decisions and will affect their loyalty to a brand. To garner and maintain that brand loyalty, companies will have to examine their manufacturing and shipping processes for carbon efficiencies. Manufacturers who do nothing to reduce their greenhouse gas emissions will find themselves losing market share, while low-carbon manufacturers will have a competitive advantage and become more successful.

Low-carbon manufacturers will gain a competitive advantage in the marketplace due to an increase in demand for low-carbon products by consumers who seek a low-carbon society. This focus on low-carbon manufacturing will drive manufacturers to use low-carbon materials and components and adjust their manufacturing processes and operations to be more carbon neutral.

As the market demand for low-carbon products increases, ICEMAN will give low-carbon manufacturers a competitive advantage. Companies that produce goods with low-carbon emissions will have a competitive advantage over those that do not. It will be devastating for companies that have a very high carbon footprint. They will have to step up their efforts to reduce their greenhouse gas emissions.

If a corporation is manufacturing a product and it's not selling as well as they would like, they will start adjusting various attributes. With ICEMAN, Carbon Factor Index value will become an important attribute that companies can adjust. Having a high CFI will create a substantial competitive advantage for a product. Companies that recognize the advantage will seek ways to increase the Carbon Factor Index values of their products—which they can only do by reducing and offsetting their carbon emissions. This would allow market forces to contribute to reductions in greenhouse gas emissions.

Once again, rather than any kind of government mandate or regulation, the demand coming from the consumer is what will help drive manufacturers to produce low-carbon-footprint products. The government won't need to regulate it, because companies will be driven to regulate themselves in order to remain competitive in the marketplace.

Market forces are stronger than any mandate that any government or coalition of governments could make. We can have a greater impact on reducing the global carbon footprint by allowing market forces to dictate the survival of the fittest—the fittest being the companies with the lowest carbon footprint.

THE POSITIVE FEEDBACK LOOP

We can begin to realize the potential effect ICEMAN may have on climate change when we consider the impact of competitive advantage.

Manufacturers will face the pressure to reduce their carbon footprint as the Carbon Factor Index casts a spotlight on manufacturers' greenhouse gas emissions by representing it on their products. Low-carbon manufacturers will have a competitive advantage and become more successful. Manufacturers that do nothing to reduce their emissions will find themselves losing market share.

A focus on low-carbon manufacturing will attract low-carbon materials and components. It will encourage manufacturers to examine carbon efficiencies in their manufacturing processes. Manufacturers will be incentivized to rebuild their plants to incorporate low-carbon systems and infrastructure to enhance their product's Carbon Factor Index value. Older facilities will have to be adapted to reduce greenhouse gas emissions. And new plants will have an incentive to incorporate low-carbon systems and infrastructure that will enhance their products' Carbon Factor Index value.

To use our earlier example of concrete clinker manufacturing, a corporation would be incentivized to use concrete that is made using the dry kiln method, which is the less carbon-intense manufacturing process. This would then incentivize the concrete factories to update their methods, operations, processes, and facilities to use that method, thereby reducing their carbon footprint.

It's a domino effect—a positive feedback loop.

The Impact on Infrastructure

ICEMAN can reach into every level of the manufacturing process, meaning it can have an impact on every stage of the creation of a product. ICEMAN will impact every product that is manufactured, every raw material, every component, every process along the supply chain, all the way back to when the natural resource is taken out of the ground. It impacts everything.

LCA is the accounting of greenhouse gas emissions during a product's entire life cycle, from the gathering of raw materials to disposal at the end of a product's life. Many products claim lower levels of emissions, but proper analysis using the Carbon Factor Index may reveal that some parts of a product's life cycle may actually compound emissions.

Because ICEMAN measures greenhouse gasses emitted throughout the manufacturing process, it gives businesses the ability to cut emissions at different stages of that process, targeting specific operations, components, or materials to reduce their overall carbon footprint.

For example, ICEMAN incentivizes manufacturers to implement renewable energy by installing energy-saving measures like solar panels at their facilities—or by relocating to an area with a low-carbon-footprint electric grid. Simply having a manufacturing facility located in a low-carbon-footprint electric grid will give a product a competitive advantage.

Take aluminum. The mining and smelting processes for metals often produce large carbon footprints. Aluminum has the highest carbon footprint, and unlike cement, there is only one method to create it: by consuming great quantities of electricity, since smelting aluminum is done by electrodes. However, an aluminum producer can move their smelter to a grid that has renewable electricity. Because the smelting process for aluminum uses a high volume of electricity, being on a renewable grid would have an impact on the carbon footprint of the aluminum produced by that smelter.

Alcoa's aluminum smelter in Massena, New York, is one of the largest in the world. However, it was built in the Massena area so that it could use the renewable hydroelectric power from the St. Lawrence Seaway. Alcoa made a business decision to locate its smelter in the Massena electric grid because the cost of hydroelectric power is significantly less than in a fossil fuel grid. When ICEMAN is implemented, that business decision becomes a no-brainer. Because it uses zero-carbon hydroelectricity from the St. Lawrence Seaway, the aluminum produced at the plant has a lower carbon footprint. When this is certified with a Carbon Factor Index value, it will have a competitive advantage in the marketplace.

Once ICEMAN is adopted, it will support and drive the whole movement of building renewable energy into our infrastructure. The movement is already happening with plans to build offshore wind farms along the eastern seaboard; ICEMAN will encourage the industry to seek out places where they can take advantage of that renewable energy. Adopting ICEMAN will attract businesses to low-carbon electric grids, incentivizing utilities to lower the carbon emissions on their grid by adopting renewable energy like offshore wind, utility-grade solar farms, or hydroelectric power. Utilities will have to come up to speed; otherwise, corporations and manufacturers are going to move to states or regions with renewable energy grids. Any state that is dragging its feet on adopting renewable energy will be left behind.

Say there is a manufacturer whose plant is at the end of its life cycle. They're looking to build a new plant. They are going to build that plant where it will be most beneficial to the company—and that may not be where the plant is currently. This is what destroyed Detroit: when the automobile plants closed, the corporations moved their operations to areas where they could work more cheaply. And Detroit is not the only place where that has happened. That is the power of market forces.

If states and cities want to prevent that from happening, they will need to invest in renewable energy infrastructure. Reducing the greenhouse gas emissions of the electric grid will reduce the embodied emissions of the manufacturers that use those grids. That is what will keep companies and manufacturers in the area. And it is what will bring companies and manufacturers to new areas—or back to areas they'd previously abandoned. If Detroit invested in renewable energy, manufacturers would move back and revitalize the city.

Once again, it creates a positive feedback loop.

A GRASSROOTS MOVEMENT

ICEMAN will work on a grassroots level. How do I know? Because the green movement is already working on a grassroots level. Individuals are taking it upon themselves to be more environmentally responsible. They are recycling and putting solar panels on their houses. But it doesn't stop with individuals. The grassroots have already grown from the individual level up to the level of towns and municipalities. I have seen this in action in my own town of East Hampton.

I am the past chair of the Town of East Hampton Energy Sustainability Committee. Under my leadership, the town adopted our Comprehensive Energy Vision and adopted the goal of being 100 percent renewable in electricity, transportation, and heating fuels. Other towns and municipalities in California had made this commitment, and we learned a great deal from them—but we were the first town on the East Coast to make such a commitment.

When we made this goal in East Hampton, Southampton soon followed with their own goal of 100 percent renewable electricity. But it didn't stop there. It wasn't long before the governor of New York was creating a goal for the state. Andrew Cuomo and his brother come out and summer in the Hamptons, so when he saw East Hampton and Southampton make these goals, he decided the state should do the same. He made a commitment for the State of New York to have 70 percent renewable electricity by 2030, and for the electric sector to be emissions-free by 2040.

The goals set by East Hampton and Southampton have spurred multiple green infrastructure projects. There are currently about six or eight projects in various stages of approval. We brought in Deepwater Wind, which secured some offshore leases from the federal government. Deepwater Wind later sold their leases to Ørsted, the largest energy company in Denmark, and one of the largest installers of offshore wind farms in the world. We now have one of the largest offshore wind projects in the US, and there are many more in various stages of approval and are even larger.

There are some additional places off the south coast of Long Island where New York has given out leases to companies to build wind farms. One of these companies is Statoil, now called Equinor, a Norwegian energy company that is reinvesting its oil revenues into developing offshore wind. They are developing a farm off the south shore of Long Island.

We now have one of the largest concentrations of offshore wind farms right off the coast of Long Island, generating energy that will replace the coal- and oil-fired plants we have now. The addition of fifteen wind turbines for the east end of Long Island was also approved by the state of New York. Between those wind turbines and the Solarize East Hampton campaign, we will be able to meet our goal of having 100 percent renewable energy in our electrical grids.

And it all started because our community set this goal. This is the exact kind of chain reaction grassroots movements can create. A municipality sees what the neighboring municipality is doing and says, "We want to do that too." Once it starts in one place, it spreads, and then suddenly you have many municipalities throughout the state making these goals. That then puts pressure on the state government to make statewide goals. And once states start making statewide goals, it puts pressure on the federal government to make national goals.

There are now at least four different wind farm developers building wind farms off the East Coast to supply electric utilities from the Boston area to the Carolinas. According to Go 100%, a group that tracks 100 percent renewable energy goals locally and globally, there are currently thirty-seven 100 percent renewable projects happening across the United States—from companies like Whole Foods and Google, to towns like East Hampton, to large cities like San Diego, Boulder, and Salt Lake City. Around the world, whole nations—including Denmark, Costa Rica, Iceland, New Zealand, and more—are pledging to go 100 percent renewable.

What makes this possible is the development of renewable energy on a grid scale. Renewable energy is replacing the old oil- and coal-fired electricity plants, which are at the end of their life cycle. On the east end of Long Island, we evaluated the cost of putting in a newer power plant versus the cost of developing renewable offshore wind energy. The offshore wind farms ended up being the cheapest out of all the energy-producing technologies. There is already a lot of manufacturing in the New York tri-state area; renewable grids will draw even more.

Even if a community is not located on a renewable grid, they can still become 100 percent renewable, through initiatives like Community Choice Aggregation (CCA), which has been adopted in Westchester County, a bedroom community north of New York City. Westchester County wasn't the first municipality in the country to develop this concept, but it was the first one in New York.

In CCA, a community or municipality brings together all of its energy users to form a buying cooperative to buy electricity. Say there are a hundred thousand people using a million kilowatts per hour. Instead of being beholden to their local utility, the CCA leverages the combined buying power of all of those electricity users to put out a request for proposal to purchase electricity from any source they desire.

Electricity travels on grids. There are local grids, state grids, and the national grid. Electrons can travel a long distance, but we can't mark the electrons and direct them where to go. So it is possible to get electricity from a wind farm in the Midwest or a solar farm in the Southwest or a hydroelectric plant in Niagara or the St. Lawrence Seaway—whether you are geographically close to those sources or not, through a mechanism developed known as Renewable Energy Credits, or RECs. This is a mechanism much like carbon offsets: the purchase of RECs justifies the electricity as renewable even when the electrons used were created by oil-powered generators.

A community can say, "We want all of our electricity to be renewable." As a municipality, they might not have the ability to build a renewable grid or generate enough renewable energy to fulfill the needs of the community. If they adopt a CCA, they can create a 100 percent renewable grid by purchasing 100 percent renewable energy from sources across the country.

In fact, you can even become a 100 percent renewable energy user as an individual, through renewable energy credits. If someone builds a solar farm, for every kilowatt of energy it produces, they earn a renewable energy credit, which is almost identical to a carbon offset credit. Like carbon offset credits, renewable energy credits are sold on a marketplace. If I, as an individual, want all the electricity in my house to be 100 percent renewable, I can purchase renewable energy credits off the marketplace equal to the amount of energy I use. Now, I'm 100 percent renewable—the same way you can be carbon neutral through a combination of mitigation and purchasing carbon offsets.

Today there is sufficient technology to provide all the world's energy needs with renewable energy. There are land-based technologies, like wind turbines, as well as offshore wind turbines. There are solar farms, like they have in California to take advantage of the endless sunshine, and in the desert, where the irradiance from the sunlight and its reflection off the sand is tremendous, producing a huge amount of electricity very efficiently. All it takes is individuals, communities, and corporations to all commit to using renewable energy. ICEMAN will help make that happen, because it will justify implementing these technologies.

GOOD FOR BUSINESS

I hope by now we've made it clear that there's a competitive advantage to adopting ICEMAN. But that's not the only reason ICEMAN is good for business. There is also a financial benefit for a company's bottom line. ICEMAN can help companies become more competitive not just in terms of their carbon footprint but also in terms of cost.

ICEMAN is free; it won't cost anything to be included in the index. We'll do it automatically, and it will be up on the website. You don't have to pay for the certification or the certified label. ICEMAN won't increase your expenses or increase the cost of your product. In fact, it will save you money. Making the adjustments necessary to earn a higher CFI value reduces the cost of manufacturing products due to the energy-saving measures companies will need to take to reduce their carbon footprint.

The fact is: it pays to adopt renewable energy. There is already a strong push for "green" products because of their cost-saving benefits. Solar is leading the way in terms of saving energy while saving money. Solar panels take about seven to eight years to make up their cost in electrical savings; the life span of solar panels is thirty years, so you get twenty-three years of free electricity.

Hydroelectricity is also really cheap. I worked on a project in Massena, and we bought, worked, and lived on a family farm there.

Our electricity prices were three cents per kilowatt hour—compared to twenty-two cents per kilowatt hour in East Hampton. Cost plus earning a higher CFI value will give a manufacturer double the reason to gravitate toward a renewable grid.

With government mandates and regulations like cap and trade, corporations will usually just push the additional costs down to the consumer, raising the price of the product. ICEMAN will have the opposite effect. It will actually save companies money.

There will be no mandate on what companies do with those savings. The reduction of costs can go toward the bottom line of the corporation, making them more profitable, and in turn they can pass that profit on to their shareholders. Or they can use it to lower the cost of their products for consumers, to become even more competitive. Corporations can decide how they want to distribute the saved cost, based on the competition in the market. You can even take the money you save by investing in renewable energy for your company and put those savings toward purchasing offsets to raise your CFI value even more and give yourself even more of an advantage in the marketplace.

It's a no-brainer. You have improved corporate image. You have increased competitive advantage. Those both come from the consumer side, from people wanting to see products that have a lower carbon footprint. From the business side, you have the benefit to the bottom line, with savings that can then be passed on to shareholders and/or consumers.

GOOD FOR THE ECONOMY

Beyond the individual business level, ICEMAN is also beneficial to the economy as a whole—on a state level and a national level. Clean energy infrastructure directly creates jobs. Building solar panels, installing and maintaining wind farms and solar farms—they are all job creators. But it goes beyond that. As we talked about, when a community—a town, a region, a state—invests in renewable infrastructure, it will draw manufacturers and companies to that area. When manufacturers and companies move into an area, they bring jobs with them.

Jobs and economic development follow industry, and industry follows the technology it needs to be successful. When railroads were built across America, towns followed the railroad. At the time, the railroad was the height of technology. So wherever the railroad went, industry followed. Industry needed transportation; it needed arteries with which to move products. So it went where the railroad was.

JOBS AND ECONOMIC DEVELOPMENT FOLLOW INDUSTRY, AND INDUSTRY FOLLOWS THE TECHNOLOGY IT NEEDS TO BE SUCCESSFUL.

Today, transportation is pretty much taken care of. You can get from anywhere to anywhere with relative ease. Today, the technology that will draw industry is renewable energy infrastructure. Just as transportation played a critical role in developing industry in the past, today renewable energy will drive industry.

COMPLEMENTARY APPROACHES

Although ICEMAN can complement a government's cap-and-trade approach to curbing carbon emissions, it is a radically different approach. Cap and trade is a top-down system that accounts for the emissions of the largest industries and biggest greenhouse gas emitters. Over a period of time, the cap-and-trade mechanism works downward to cap greenhouse gas emissions of smaller industries and emitters.

ICEMAN works from the bottom up. It incorporates the calculation of the carbon footprint of small business operations—including the small businesses that make components and ship goods. Since small businesses represent over 99 percent of all businesses in the US and employ nearly half of all US employees,[18] ICEMAN has the potential to engage the interest of a much larger audience and could curb emissions from a wider range of businesses than cap and trade can.

The ICEMAN mechanism and cap and trade are not at odds with each other, nor do they compete against each other. They work in harmony as different but complementary approaches. Properly implemented, the ICEMAN mechanism is designed to exist harmoniously with other policies. For example, cap and trade focuses on the carbon emissions of the manufacturer alone, while ICEMAN includes the carbon emissions of the product *and* the manufacturer combined, providing a more comprehensive solution to carbon emissions as a whole.

Especially in the United States, government regulations tend to have negative connotations, whereas mechanisms that stimulate corporate growth and fiscal health are generally considered positive. A mechanism that harnesses positive market forces and that has the ability to work in harmony with mandated policies can bring together the best of both worlds.

Mandated systems need to work and coexist in harmony with market-driven mechanisms such as ICEMAN. This is the only way we can slow down the exponential growth and eventually reduce greenhouse gas emissions.

---

18 JPMorgan Chase Institute, "Small Economic Activity," accessed May 2, 2021, https://www.jpmorganchase.com/institute/research/small-business/small-business-dashboard/economic-activity.

CHAPTER 6

ICEMAN: FOR THIS MOMENT, AND FOR THE FUTURE

I've shown ICEMAN to both Republican and Democratic legislators, and they have all been hugely supportive. I have brought ICEMAN before several members of Congress and of the Republican Senate Committee on the Environment with no opposition. My GOP Congressman has already honored me and ICEMAN, placing me at the seat of honor at a Long Island GOP breakfast. I've been in rooms with the upper echelons of the Republican Party, and everyone has offered only praise for this idea.

I've met with Senators, Congressmen, and senior regulators in the EPA and DOE. I sent them my executive summary, then met and sat down with them face-to-face. There's been absolutely no resistance at all. Everyone I've talked to has offered total support. Republican, Democrat, legislator, and regulator, no one has any opposition.

I had legislators help direct me to the right agencies to help implement ICEMAN. They introduced me to people high up in the

EPA and the DOE. I met with them, and we discussed how we can form a public-private partnership with the government.

The senior policy analyst and foreign affairs guy from the DOE loved the concept. "I want to let you know," he said, right off the bat, "I read your executive summary cover to cover, every word of it. And I absolutely love it." The person from the EPA explained to me that they have a structure within the EPA to form a public-private partnership with me to implement ICEMAN.

No legislator or department official I've showed ICEMAN to has been able to shoot any holes in it.

So why hasn't ICEMAN already been implemented?

The answer is simple: political capital. Until now, there simply has not been the political capital to fund the project, in either the DOE or the EPA. Nothing can move forward until funding and the allocation of personnel is approved by Congress. There must be a legislative act to fund the EPA and possibly the DOE to work on this project and assign personnel to it.

Now, the time is right.

Before, there wasn't the groundswell from the grassroots needed to get the government to respond. Now, we have reached a tipping point. The moment has come when there is enough of a groundswell of support for saving the environment that legislative action can be taken.

A NEW FOCUS ON CLIMATE

The Biden Administration has clearly and forcefully made environmental policy one of its top priorities. Fighting climate change and creating green jobs are at the forefront of the agenda. Within hours of assuming the office of president, Joe Biden signed an executive order for the United States to rejoin the Paris Agreement.

A week later, on January 27, 2020, Biden signed an Executive Order on Tackling the Climate Crisis at Home and Abroad. This Executive Order explicitly puts the climate crisis at the center of United States Foreign Policy and National Security. The Executive Order also pledges a government-wide approach to the climate crisis, citing "opportunities to create well-paying union jobs to build a modern and sustainable infrastructure, deliver an equitable, clean energy future, and put the United States on a path to achieve net-zero emissions, economy-wide, by no later than 2050."[19] The order also establishes a White House Office of Domestic Climate Policy and a National Climate Task Force, chaired by the National Climate Advisor and made up of numerous high-level departmental secretaries, including the Secretary of Defense, the Secretary of the Treasury, and the Attorney General.

On April 22, 2021—Earth Day—President Biden hosted an Earth Day Climate Summit with forty world leaders. At this summit, Mr. Biden formally pledged that the United States would cut its greenhouse gas emissions in half by the end of the decade, setting an economy-wide emissions target of a 50–52 percent reduction below 2005 emissions levels by 2030. The United States also restated its commitment to transitioning to net-zero emissions no later than 2050.

IF YOU DON'T GET ON BOARD, YOU'RE GOING TO BE LEFT BEHIND.

These policies are long overdue—and the time is right for ICEMAN. Now, there is the political capital to fund this project. This

19 The White House, "Executive Order on Tackling the Climate Crisis at Home and Abroad," January 27, 2021, https://www.whitehouse.gov/briefing-room/presidential-actions/2021/01/27/executive-order-on-tackling-the-climate-crisis-at-home-and-abroad/.

administration is going to continue to push hard on climate change. If you don't get on board, you're going to be left behind.

ICEMAN can become a part of this administration's progressive climate agenda. In fact, ICEMAN can be the missing piece of the puzzle that makes the Green New Deal being proposed by legislators like Representative Alexandria Ocasio-Cortez really work.

The greatest concern about the current Green New Deal legislation being proposed is the economic impact. Critics of the Green New Deal are concerned that it will be an economic catastrophe. Implementing ICEMAN will alleviate that criticism and fear from the other side. If the Green New Deal incorporated ICEMAN, it would eliminate the opposition that predicts economic disaster. In fact, I believe it would just about guarantee that the Green New Deal gets passed. ICEMAN would be a benefit to the Green New Deal programs while benefiting the economy at the same time. It becomes a win-win situation for everyone—and for the planet.

EVERYBODY'S GETTING ON BOARD

Even the most unlikely of corporations are getting on board with going carbon neutral—and these pledges began even before the Biden Administration took office and started pushing its progressive climate agenda.

In December of 2019, Spain-based oil company Repsol pledged to become carbon neutral by 2050—the first major oil company to make such a pledge. Repsol plans to produce biofuels, which is a renewable energy source from biomass, not fossil fuels. This is a great example of adaptation resulting from the goals set by the Paris Agreement. Others will then follow—and already have—because they have seen how Repsol's goal gave Repsol a competitive advantage and improved the corporate image. Other oil companies don't want to be outdone.

Occidental Petroleum, one of the United States' largest oil drillers and the fifth-largest US oil company by market valuation, declared its intentions to be carbon neutral by 2040. Naysayers say it is impossible to offset the burning of the oil produced by Occidental—the emissions created by cars, busses, airplanes, and other vehicles and machinery that run using Occidental's oil.

But that is not what is meant by the business being carbon neutral. As we discussed earlier, we are talking about a cradle-to-gate approach, not taking into account the downstream emissions of the oil. The downstream emissions are accounted for by the carbon reporting of those who burn the oil, not those who manufacture it. For example, the burning of oil or gas is accounted for by the business that uses that oil for heating, transporting goods, etc. The burning of oil and gas for the production of electricity is accounted for by the utility generating electric. The burning of jet fuel, diesel, and gas is included in the corporation's upstream Scope 3 emissions used in travel. The rule of thumb is carbon emissions are accounted for only once.

Occidental plans to capture its CO_2 emissions and inject that CO_2 into the ground as part of a process called enhanced oil recovery. This will prevent that CO_2 from entering the atmosphere.[20] This is a legitimate carbon offset project, and Occidental will be entitled to take those carbon credits. The fact that even so-called experts don't fully grasp how this works shows we need a lot of education to bring everyone up to speed and singing from the same hymnal. This is why ICEMAN uses the WRI Greenhouse Gas Protocol: because it

20 Matt Egan, "One of America's Biggest Oil Companies Wants to Be 'Carbon Neutral'—Eventually," CNN Business, March 20, 2019, https://www.cnn.com/2019/03/20/investing/occidental-carbon-neutral-oil-shale/index.html.

is simple to understand, is a widely accepted protocol, and can be a go-to reference for how the entire process works.

In February of 2021, Shell, Europe's biggest oil firm, announced its goal to have net-zero emissions for both itself and its products used by customers by 2050. Shell wants to expand biofuel and hydrogen production, both renewable energy sources.

Investors themselves are becoming drivers for companies to decarbonize, as activist investors force corporations to address climate change concerns. As of June 2021, the activist investor Engine No. 1 holds two seats on the board of Exxon Mobil, with a third seat looking likely. With its seats on the board—which will represent a quarter of the board, if it gains the third seat—Engine No. 1 is pressuring Exxon Mobil to become carbon neutral by 2050.[21]

These companies are voluntarily setting these goals. No government is mandating that they do so; they are doing it of their own volition. This demonstrates the huge corporate pressure to adapt. We wish to add to that pressure by implementing ICEMAN and letting the market forces of competitive advantage go to work.

GLOBAL IMPACT

What the United States does, Europe will follow. For some time now, the United States has had a bad reputation when it comes to reducing greenhouse gas emissions. When I presented the concept of ICEMAN at a Zero Emissions conference in Oslo, after the presentation someone stood up and said, "I can't believe something that advanced came out of America!"

21 Christopher M. Matthews, "Activist Likely to Gain Third Seat on Exxon Board," *Wall Street Journal*, June 2, 2021, sec. Business, https://www.wsj.com/articles/activist-likely-to-gain-third-seat-on-exxon-board-11622664757.

"That's because I was born in Norway," I replied.

ICEMAN offers a chance for us to turn things around, to be a leader in reducing our carbon footprint. We can show the world how this can be done in a free market economy. If the United States adopts ICEMAN, it will put America back on the map as a world leader in reducing carbon footprints. And what the US does, Europe will follow, and the rest of the world will too.

If implemented in the United States, the impact of ICEMAN will extend far beyond this country. Manufacturers worldwide will have to step up to adopt renewable energy and reduce their carbon footprints in order to be competitive in the world market. Market forces alone will pressure other countries to invest in renewable energy infrastructure. As manufacturers gravitate toward regions with low-carbon infrastructure, companies that want to attract business and industry will have to step up. It is possible that entire countries will attract industry based on their low-carbon infrastructure.

Industries may be drawn to countries like Norway, which has a 99 percent hydroelectric grid. Right from the beginning, Norway built its infrastructure to rely on hydroelectric power. Norway has a lot of mountains, with a lot of waterfalls, so they were able to use the falling water to power generators. At night, there's less demand for electricity—but water is still running off the mountain. So they use that extra electricity to pump the water back uphill, so it can be reused the next day to generate more electricity.

Currently, Norway doesn't have a lot of industry. There aren't many manufacturers who export from Norway. In fact, Norway's biggest export is oil. The reason Norway isn't competitive in the world manufacturing market is because labor costs are relatively high. However, once the marketplace really starts to consider carbon footprints, Norway is going to become much more competitive, due to its 99 percent renewable grid. Even if companies are paying a little bit more for labor, it will be exceeded by the overall competitive advantage they'll gain in the marketplace as a result of ICEMAN.

Countries with high-carbon infrastructure, on the other hand, will come under pressure to rebuild their infrastructure to be more carbon neutral, to prevent industry from moving away. Meanwhile, developing countries will have an incentive to build low-carbon infrastructure to attract industry to help develop their country. When I was invited to speak in Myanmar, the term "leapfrogging" was used, meaning building a renewable infrastructure without making the same mistakes industrialized countries made in their beginnings. And countries that do nothing or just maintain the status quo may find themselves falling behind the rest of the world.

The US could encourage this by requiring every product imported into the country to have a CFI value on it. They wouldn't have to mandate products that have a CFI above a certain level; market forces would take care of that. All they would have to mandate would be that the product simply has to have an Index value.

When I presented ICEMAN to the Norwegian parliament, I met with the gentleman who is now the Minister of Transportation, who at the time sat on the parliament's standing committee on the environment.

"This is a great idea," he said to me. "We are importing products from China. We don't know the carbon footprint of these products. If we know what the CFI [values] of the products are, we can establish a tariff system on imported products below a certain CFI value."

The idea of the Carbon Factor Index being used for tariffs had never even occurred to me. The fact is I don't particularly care how governments use the Carbon Factor Index. I'm not a legislator or policy maker; I leave that in the hands of the experts. However, I could foresee that, if the market forces don't kick in as strongly as we anticipate globally, countries could put tariffs on imports that don't meet a certain Carbon Factor Index level. That would put pressure on those countries, such as China, to conform to carbon-level standards.

A New Global Protocol

ICEMAN could also be incorporated into future international climate agreements. History has shown us that mandates don't work. From the Paris Agreement all the way back to the first Earth Summit in Rio in 1992, we have tried the route of mandating reductions in carbon footprint.

At the Paris Agreement, they finally realized handing down mandates wasn't working. Instead, they asked countries to set their own goals. This may seem like a good, equitable idea, but if you look at the actual commitments made by countries as part of the Paris agreement, you'll realize that it doesn't actually do a whole lot for reducing the overall global carbon footprint in a timely fashion to avoid catastrophic, irreversible environmental harm.

In accordance with the Paris Agreement, China made a statement saying they will start reducing their carbon footprint—*beginning* in 2030. This means that from now until 2030, their carbon footprint will continue to grow, counteracting the reductions happening in other countries. In fact, China's carbon emissions are increasing so exponentially that in the next nine years, they may completely wipe out the reduction in the global carbon footprint made by the rest of the world.

Annual CO_2 Emissions by Country

Carbon dioxide (CO_2) emissions from the burning of fossil fuels for energy and cement production. Land use change is not included.

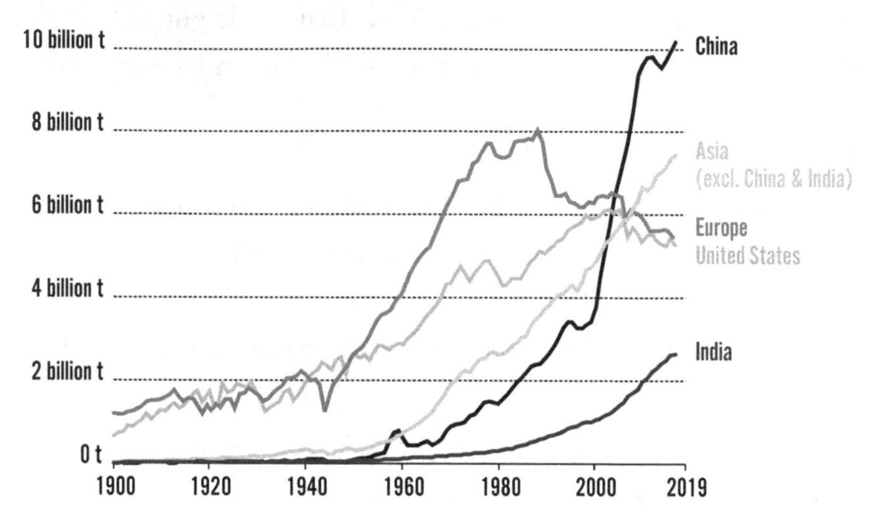

Note: CO_2 emissions are measured on a production basis, meaning they do not correct for emissions embedded in traded goods.

Source: Global Carbon Project; Carbon Dioxide Information Analysis Centre (CDIAC)

All of our commitments don't mean much if countries like India and China aren't making the same commitments. However, it doesn't do much good to just point fingers and shake our heads. What can we actually do to help the situation?

ICEMAN will be able to help solve this dilemma of countries with runaway carbon emissions. The federal government can implement a regulation requiring a Carbon Factor Index number on all imports entering the US. This will have a direct impact on foreign countries' manufacturers who export to the US market. It will incentivize them to reduce their carbon footprint, thus lowering the total carbon emissions in that country and the world.

For example, a manufacturer in China consuming electricity from a coal-fired grid will be pressured by market forces to install renewable energy to be able to compete against US manufacturers that are on a renewable grid or that have already installed renewable electricity. It's not too far-fetched to say when the US adopts ICEMAN, it may have a global impact on reducing carbon emissions worldwide.

Even without a regulation from the federal government, ICEMAN will influence manufacturers in China who import to the United States. If consumers in the US start buying products with higher CFI values, manufacturers in China will have to reduce their carbon footprints and get better CFI values in order to remain competitive in the marketplace. I find this fascinating; since lower costs won't be China's only competitive advantage, a lower carbon footprint will be an added attribute for the consumer to consider.

Tracking Global Goals

ICEMAN can also be of great help in terms of tracking the long-term goals set in international climate agreements. The concept of carbon neutrality is popular because the goal is simple to comprehend: offsetting the carbon emissions of a certain activity to the point where there is no impact on increasing greenhouse gasses. But how do you measure progress toward carbon neutrality over the long term? If a company or a country has made a commitment to be carbon neutral by, say, 2050, as Norway did, how do you comprehend, let alone track, that large a commitment over that long a period of time?

However, when you try to apply that concept to every activity that goes into creating a single product or service, and then when you expand that to entire industries, entire countries, entire economies, it gets very complex very quickly. It is incredibly difficult to comprehend the vast number of intricate moving pieces that need to be assessed and adjusted to meet that commitment over a thiry-year period of time.

With a comprehensive and standardized methodology to measure the percentage of carbon neutrality of any given product or service, measuring progress becomes much easier. ICEMAN can calculate those percent values, which can be reported to show how the country is progressing toward its long-term goal.

THE FUTURE OF ICEMAN, AND THE FUTURE WITH ICEMAN

While ICEMAN is currently focused on manufacturing, it can potentially be expanded across every industry. ICEMAN already works for the food-processing industry; the emissions from the manufacturing process for foods can be measured and calculated the same way as any other manufacturing process. Right next to the organic label on your cereal box would be the CFI label, letting people know how green the cereal is. But we can expand ICEMAN beyond that, right back to the farm. We can work with farmers to develop a system that works for the agricultural industry, creating an index for farm goods.

We can do a Carbon Factor Index for automobiles. Today's electric vehicles are actually a perfect illustration of why ICEMAN is needed across all industries. Electric vehicles are touted as green, because they produce zero emissions when you drive them. However, the carbon emissions created by the process of manufacturing the batteries for these cars is enormous. Mining the lithium needed for the batteries has a tremendous impact on our environment.

Then there is the carbon footprint of the electricity used to charge electric vehicles. In fact, the carbon footprint of the manufacturing process for an electric vehicle plus the carbon footprint of the electricity used to charge the vehicle is almost the same as the carbon footprint of driving a gas-powered vehicle. They actually have just about the same impact on the environment.

I have done studies on the carbon footprint of an electric vehicle versus the same vehicle powered by gas. I compared a Nissan Leaf and a Nissan Versa—the same car, same platform, but one is electric, and one is gas. I looked up the gas mileage of the Versa, and through the EPA I looked up the carbon footprint of burning a gallon of gas. Then, I calculated the carbon footprint of charging the battery of an electric vehicle. If you are charging the vehicle on a fossil fuel grid, you are causing more greenhouse gas emissions than if you drove a gas-powered car.

This is counterintuitive because companies advertise electric vehicles as having no tailpipe emissions—which the average consumer hears as zero emissions overall. But this doesn't take into account the emissions caused by the manufacturing process or the electricity used to charge the car.

The automobile industry has been working on gas engine efficiency since the 1970s. Cars now have catalytic converters, which scrub the exhaust, converting carbon monoxide to carbon sulfite, which is not as dangerous. The pollution coming out of gas engine cars today is drastically reduced from what it was in the 1970s.

Once all of this was taken into account, the Nissan Versa actually had a smaller carbon footprint than the Nissan Leaf.

Now, if you could charge your electric vehicle on a renewable grid or use solar panels on your house, then an electric vehicle would surpass a gas-powered vehicle. Which brings us back to our positive feedback loop: if the grid in a region goes renewable, having an electric vehicle makes sense. In fact, it's much better than having a gas vehicle. So as grids go renewable, more people will get electric vehicles, reducing the number of gas-fueled cars and the greenhouse gas emissions they create. And as more people learn the true carbon footprint of an electric vehicle, the more they will want to live on a renewable grid, putting pressure on their communities, towns, and states to switch to renewable energy.

To illustrate this point, Norway has the world's largest electric vehicle ownership per capita by a large margin, more than three times the next highest country, Iceland.[22] Because Norway has a 99 percent hydroelectric grid, the country gave incentives to increase demand in electric vehicle ownership by an exemption of a car ownership tax, annual road tax, public parking fees, toll payments, and domestic ferry costs and were given access to bus lanes.

CONTINUING POSITIVE FEEDBACK LOOPS

As people start using green technologies like solar panels more, those technologies will continue to improve and become even more efficient and energy saving. The technologies will become more available and affordable, so more people can do things like making their houses fossil fuel–free, as I am doing.

This project was inspired by a professor from Stony Brook who teaches thermodynamics, and he came up with a concept of creating homeostasis in buildings by mimicking how our bodies maintain temperature by circulating fluids. He put that idea together with a concept that came out of Germany called TABS: Thermally Activated Building Systems. In this concept, the mass of the building itself becomes a reservoir for both heating and cooling.

The professor wrote a paper, putting the two concepts together, and sent the paper to me. "Can I come over and talk it through with you?" he asked.

We ended up talking late into the night about whether and how

---

22 "Electric car use by country," Wikipedia, accessed August 2, 2021, https://en.wikipedia.org/wiki/Electric_car_use_by_country.

these concepts could be combined. "Doc," I said, "I know how to build this. I know about certain emerging technologies I can implement in a unique way to make it work."

"Amazing," he said. "Let's do this project together. We'll look for a building at the university we can overhaul. We'll get grant funding for it and everything."

Unfortunately, we weren't able to do that. Stony Brook is a state university, and he couldn't get through the red tape. "Well," I said, "I want to put this addition on my house. Why don't we use my house? I'll fund it, and you and I will study it. It will be a living laboratory, a proof of concept."

I am currently underway with this process, and we are studying it as we go. We will be writing scholarly papers on what we are discovering. We've developed theories of how it will work; this will be the place to try and prove them.

First, I focused on making it more energy efficient. I installed solar panels and adopted energy-saving technologies. It's like an energy battery. Right now, at my home I am producing 140 percent of the electricity I consume. I have 40 percent more electricity than I need! That excess electricity is going into an energy bank. I'm using the mass of the building as an energy reservoir for both heating and cooling—that mass being concrete.

Because I am producing so much electricity, it would make sense for me to have an electric vehicle, because I could charge it without using electricity from a nonrenewable source. And because I would never need to purchase gas, and I would be using electricity I've produced myself, the car would run for free! As always, it's all connected. It all feeds into itself.

I'm living in the house right now, and some of the technologies are already active, so I know they will work. Now we need to collect the data and codify the engineering so we can replicate it. That way, the next house I build, we will be able to properly engineer the various technologies for a fossil fuel–free home.

A project like this works in conjunction with ICEMAN. We're replacing fossil fuels with renewable energy, reducing the carbon footprint of the operations stage of the house.

THE FUTURE IS BRIGHT

As you can see, once ICEMAN is implemented, its effects will snowball. One little label on products will cause monumental shifts across all industries, spreading through our infrastructure, and expanding across the entire globe. Each shift toward reducing carbon footprints will trigger more shifts, until the whole world starts moving closer and closer to decarbonize. Suddenly, a decarbonized world doesn't seem so impossibly far away, does it?

CONCLUSION

Sometimes a simple idea can change the world.

We all know that burning fossil fuels creates pollutants and that pollutants are bad for us. They poison our air and our water, impacting our health. It doesn't matter if we are Democrat or Republican. It doesn't matter if we live in the city or the country. It doesn't matter if we live in the United States, Europe, China, or Africa.

We all breathe the same air. We all drink the same water. We should all be equally concerned about the environment. We may disagree about how to handle it. But the fact that it needs to be handled is something we can and should unite around.

THE WORLD IS ON A ROAD TOWARD DESTRUCTION—BUT WE CAN STOP IT.

The world is on a road toward destruction—but we can stop it.

I believe the future is hopeful. I see it in the way the generation after mine—my daughter's generation—is leading the charge to fight climate change. When I talk to people in my generation, the Boomer generation, I have to explain exactly what climate change is. They may argue with me. They may deny climate change. I've spoken at events and been openly heckled by climate deniers from my generation.

My daughter's generation looks at us and says, "You're leaving us a world in peril. We're going to do something about it." And they're right. The sad truth is my generation screwed up a lot. And I'm taking responsibility for that. That's why I do what I do.

The upcoming generation understands that it is their responsibility to make things better. The upcoming generation understands that they need to take care of the environment, because they have to live in it, and their children will have to live in it. And they are willing to make choices to achieve a better future.

Again, organic food is a great illustration. The upcoming generation is far more educated than my generation when it comes to eating healthily and taking care of their bodies. Organic food may be slightly more expensive, but it's worth it to have healthier bodies, to live longer, to have better quality of life—and, in the long run, to actually save money on health issues and medical costs.

The same goes for the environment. Most people of the younger generation are willing to spend a little more money to effectuate that change, knowing that it will save them—save them money, save their lives, save the planet—in the long run.

The next generation—my daughter's generation—has started to open my generation's eyes. We're starting to understand that what we did was wrong. Now, we have to correct it, before handing the planet over to our children. Now, we have a chance to forward a movement to reverse all the bad things that we have done for the environment, going all the way back to the Industrial Revolution. Now, we have the chance to say, "How can we fix this? How can I as an individual take responsibility to make this planet a better place?"

A movement is made up of individual people, and the United States is all about We, the People. People have to take personal responsibility first. As people take personal responsibility, they'll inspire their friends, their neighbors, their communities, building the movement into a collective effort. As more individual people and communities get involved, it inspires even more people and communities to get involved, creating a positive feedback loop.

And I believe we will start seeing more and more people take responsibility. I think we'll see it happen more in the United States than in any other country, because we have that grassroots "We the People" ethos. We understand that we are a privileged nation, that we have rights—and that along with those rights come responsibilities.

Individually, people are getting on board with the effort to save the environment. And at some point, it will reach a critical mass, and the whole country will come on board. I believe this is happening as we speak.

What will finally tip the scales? Knowledge. Knowledge is power. By providing knowledge to consumers, ICEMAN will have a powerful impact. The consumer creates the demand. The more information an educated consumer has, the more they will determine the success or failure of manufacturers. The CFI label provides that information to the consumer, giving them the power to reduce carbon emissions.

With a CFI label on every product, we can empower the marketplace and market forces to reduce carbon emissions—faster, more effectively, and more efficiently than any mandated system.

ICEMAN has the potential to help the US exceed the commitments laid out in the Paris Agreement by a huge margin. It can help the United States become a world leader in reducing our carbon footprint and in leading other countries to reduce their carbon footprints.

ICEMAN will restore integrity to the green movement by mathematically defining green based on comprehensive measurements of greenhouse gas emissions, so consumers around the world can make fully informed purchasing choices and have a direct impact on the global carbon footprint.

Because ICEMAN benefits everyone, it doesn't matter what your view is on the causes of climate change. There's no reason *not* to implement ICEMAN. That's why it will be successful. ICEMAN brings everyone together in a win-win solution. Legislators, business owners, and consumers can all unite and participate in ICEMAN.

But all of this can only happen if we act with the urgency of now.

If you are a legislator, I am asking you to take action on implementing ICEMAN. Pass legislation requiring businesses to report their greenhouse gas emissions, regardless of size. Empower the EPA and DOE to fund and support a partnership with ICEMAN.

If you are a business owner, I am asking you to support ICEMAN, to lobby for it to be implemented and, when it is, to include CFI labels on your packaging. I am an entrepreneur with a small business. I've seen firsthand the benefits of going carbon neutral, as we did with the HGA House. I know there are others out there who will want to do the same. I'm asking you to be one of them.

If you are a consumer reading this book, you don't have to wait for the government and businesses to take action. You can contact your representatives and tell them you support this and that you want them to support it. Representatives want to know what their constituents want. It's how our political system works. Tell your representatives to implement ICEMAN.

We have to do this now, and we have to do it together. If we do, we can, quite literally, save the world.

CONTACT

For more information on the ICEMAN Carbon Factor Index and to learn how you can help decarbonize the world, please go to **carbonfactorindex.com**.

You can also reach out to Frank Dalene using the following info:

frank@carbonfactorindex.com
(631)537-1600
Carbon Factor Inc.
PO Box 1071
Wainscott, NY 11975

CPSIA information can be obtained
at www.ICGtesting.com
Printed in the USA
JSHW052328051021
19323JS00005B/35